BUILDING
FOR garde
GW00686403

Published by Bay Books, 61-69 Anzac Parade, Kensington, NSW 2033

National Library of Australia
Card number and ISBN 1 86256 270 9

The publisher wishes to thank the following for their assistance with additional photography for this book:
Australian Picture Library — pages 29, 31, 50, 61
Black and Decker — pages 24, 25
Keith Manns — pages 7, 11, 15, 16, 18, 22, 31, 42, 46, 50, 59, 60, 61, 63, 77, 85
Mote Ladders — page 26

Photography: Ray Joyce

Designed by Tony Theunissen

Typesetting by Savage Type

Printed in Singapore by Toppan Printing Co

BB88

BAY BOOKS PRACTICAL DESIGNS FOR LIVING

BUILDING FOR gardens

DECKS • PERGOLAS • GAZEBOS • PLAY STRUCTURES • FENCES

ROB & SUE WHELAN

BAY BOOKS
Sydney & London

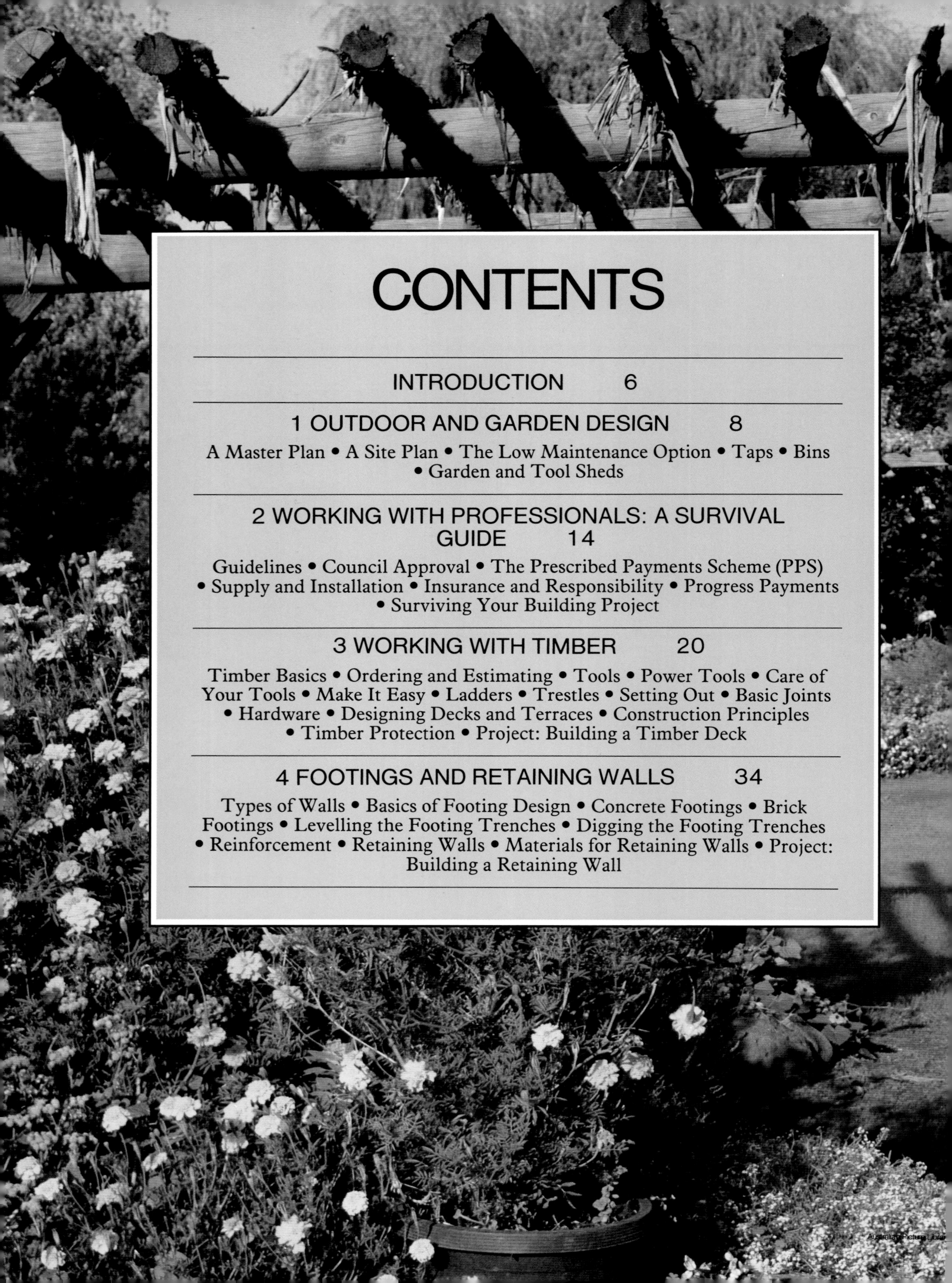

CONTENTS

Australian Picture Library

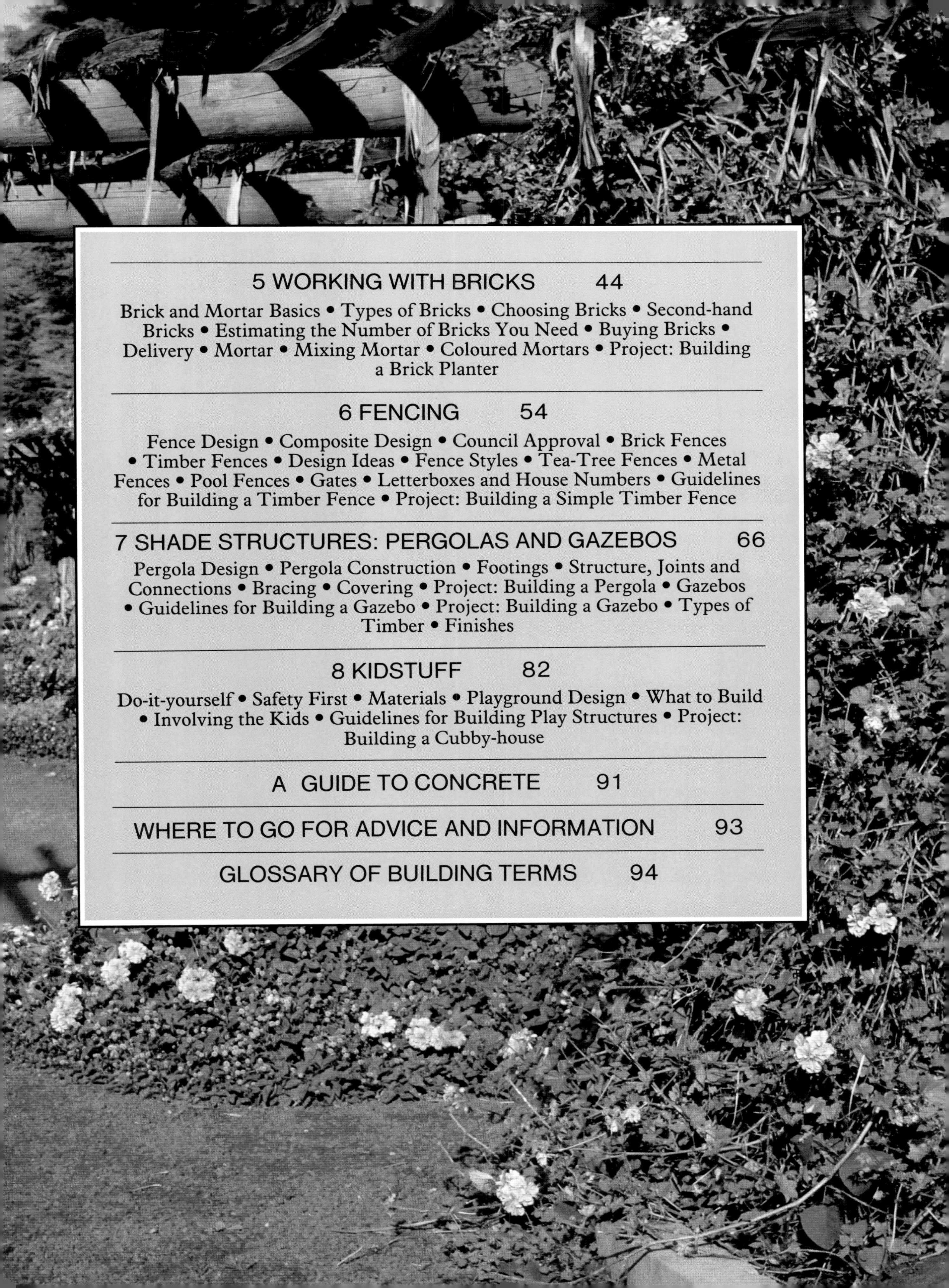

INTRODUCTION

Your garden and living areas are an important all-year-round part of your family's home life. The general design and appearance of these areas should reflect your lifestyle as much as the overall design and decoration of your house.

Designing and building your own garden structures, from walls and fences to pergolas and even play structures for your children, can be highly rewarding. Some garden design projects are quite challenging, but very few are beyond the abilities of the average homeowner.

A well-designed garden project will add value to your property, as well as provide a more attractive outdoor living space for home entertaining and family leisure time.

Given a basic tool kit, a few design hints and a working knowledge of some of the tricks of the trade, garden landscaping and structural work can be a real pleasure. Creative design and intelligent use of materials is essential, but this book is dedicated to the idea that with good design and preparation, virtually anything is possible for the amateur.

There are some jobs where it is better to employ a professional. There is a detailed 'survival' guide to dealing with the building industry which gives some of the do's and don'ts, and hints on the best approach to briefing, employing and working with specialist consultants — like architects and landscape designers, as well as tradesmen.

A glossary of building terms has been included at the back of the book for your reference. The language of the building trade is full of special trade terms that you will need to be familiar with. It's important that you can communicate what you want to the pros, and equally important that you understand what they're talking about when discussing your job.

This book is designed for do-it-yourselfers who want to learn a few trade secrets and some of the easy ways to approach home improvement projects. It will alert you to the pitfalls of home improvement projects, and how to avoid them. Building materials are too expensive to waste and even if you hire a professional you need to know what's involved in the job yourself — just to make sure that the work is done properly.

Proper planning and a knowledge of estimating and ordering materials for home improvements are critical factors in achieving your overall design plans. Much of the cost of any tradesman's quote is for labour. If you can do most or all of the work yourself, you will not only save money but also gain the satisfaction of taking on a challenge and succeeding. The experience gained on one job can be applied to the next and, as your confidence grows, there will be no limit to the scope or complexity of your projects.

1

OUTDOOR AND GARDEN DESIGN

A Master Plan • A Site Plan • The Low Maintenance Option • Taps • Bins • Garden and Tool Sheds

It's important that your garden and outdoor living areas suit your lifestyle, family needs and activities. Regardless of whether your recreational areas are mainly at the front, back or sides of your home, planning how everything fits in together, now and in the future, can be difficult.

The ever popular and constantly used backyard can be a special problem area. The classic 'problem' yard has a clothes line in the middle, a shed down the far end, a messy perimeter of plants and an ugly, battered fence! Sound familiar?

It's often difficult to know where to start . . . after all, everything needs a place — kids' toys and bikes, the lawnmower, garden tools, boxes full of collector's junk and the clothes line. Sooner or later, you'll have to rethink your outdoors to suit your family's changing needs.

Ideally, your outdoors should be an extension of your home and a focal point of summertime entertaining and family life. Home improvements can minimise maintenance and allow you more leisure-time to enjoy the results. Storage and clothes-drying areas are necessary but don't need to dominate your garden. Sheds can be designed to blend in, camouflaged by plants and screens. The latest designs in hide-away clothes lines are designed to take up the minimum amount of garden space.

Paving and carefully thought-out pathways can be used to connect different areas, and reduce problems when the ground is wet. Terracing can also be an attractive and interesting way of separating activity areas by platforms built on different levels. This is especially so if your house is built on a steep, sloping site. Pathways and a good terracing system can help prevent interior floor damage due to dirt and mud being walked in from outside. The need for adequate drainage to allow for run-off of rainwater and falls to drains and pits is also important to consider.

A Master Plan

Before you start work on any single home improvement project you must have a master plan. This plan should include all of the features you want to incorporate in your garden over a period of time. In this way any problems can be worked out before construction gets underway. A plan is also an excellent means of visualising the end result. Try to anticipate future needs by making your plan as flexible as possible.

Write down a list of the features you want to include in your garden. Rank items according to their relative importance — also write a list of 'must haves', 'like to haves' and 'don't wants'! Use these lists as a design tool to check off how well the design fits your needs.

If you're going to enlist professional help, for example an architect, drafting service or landscape gardener, your master plan and lists will be invaluable. From these a designer will be able to map out the best way of fitting it all together.

Above: This well-designed tree-house makes the most of its surroundings by blending in with both the tree and the steep sloping section it is connected to by a footbridge.

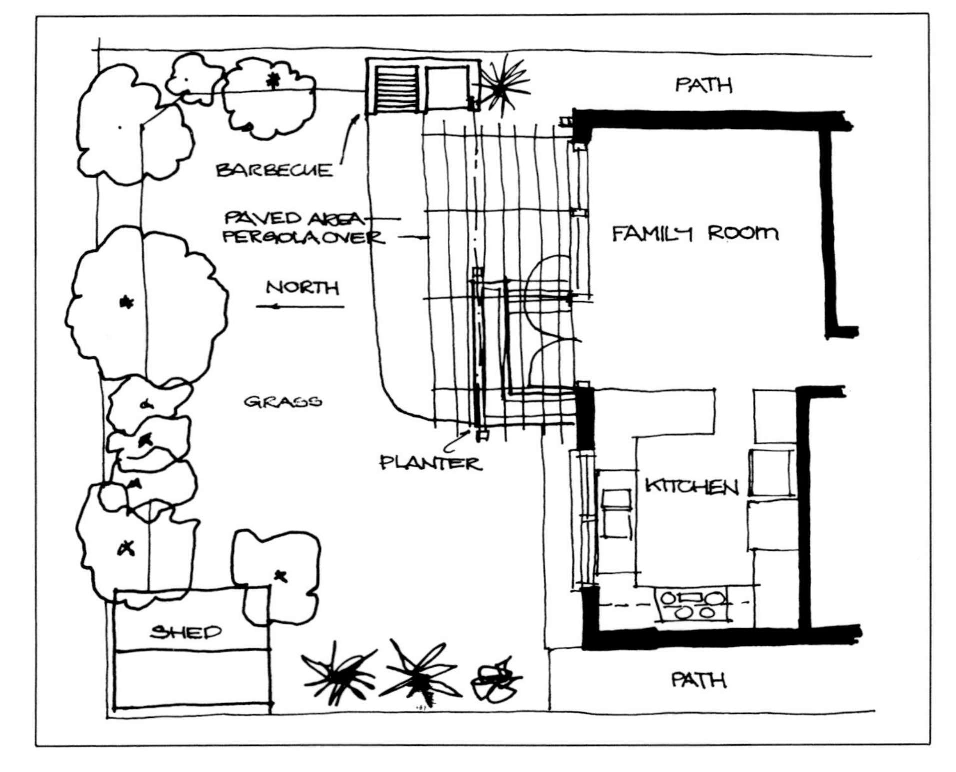

This rough concept plan for a barbecue and outdoor living area utilises its sunny north-facing position and at the same time incorporates shelter and shade with a pergola attached to the house.

Below left: This pergola creates an attractive setting for outdoor eating. Right: This log retaining wall has been designed to create a compact terraced garden area.

Above: Terracotta pots, plants in a wheelbarrow and a gazebo (*background*).

Above: A well-designed barbecue is an important part of any outdoor entertainment area.

A Site Plan

The first thing an architect or designer will do before starting work on a new job, is to draw up an accurate site plan. It's a good idea to get the plans of your house from the builder, architect or previous owner — if you don't already have them. If you can't find the plans and your house was built in the last twenty years, try to get a copy from your local council. Most councils keep microfilmed records of previous building applications. Check whether any alterations have been made to your house since the last application to council. Sometimes people make small alterations and additions to their properties without council approval.

Your solicitor might have a copy of the survey plans, done when you purchased your home. These plans will at least show site boundaries and the position of your house in relation to neighbours. Photocopy the plans and carefully measure and record the position of any significant features. When you record the position of trees, try to estimate how much they have grown and how big they'll be in five years' time. Remember this is part of your master plan for the present and the future.

Referring back to your lists of things you want to have in your outdoor area, think about where everything will go. Carefully consider different activities and how they interrelate. Do you really want the barbecue next to the compost heap? Should the clothes line go directly under that tree? These initial thoughts, lists and layouts are a great way to test all the possibilities before your final design decisions are made. Don't be afraid to sketch out ideas and to throw away the schemes which don't work. Mistakes made on paper are simple and cheap to correct, unlike those made during work in progress!

Below: A survey plan.

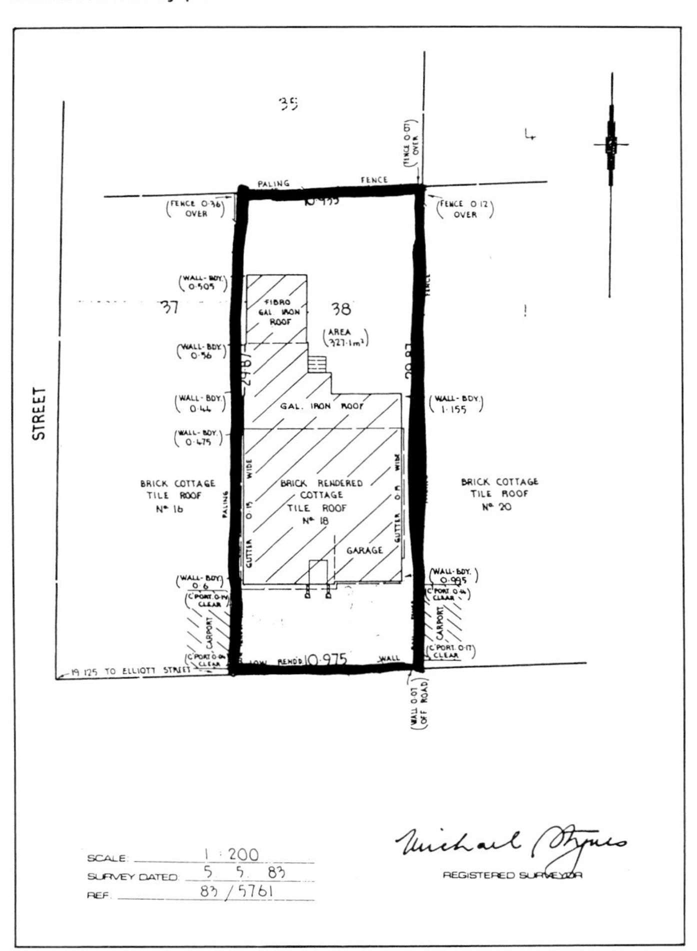

The Low Maintenance Option

Low maintenance surfaces and finishes such as brick and concrete unit paving, face brickwork, stone and slate, treated timber, stain finishes to timber and high durability paint will all help to reduce future problems. If you have to make a choice between two systems or materials, and the maintenance factor is the main difference, it's usually wiser to choose the low maintenance option.

Above: This rough face stonework acts as a transition area between two different levels.

Above: Brick unit paving is a safe, durable, low-slip surface for the pool area.

Above: Traditional trelliswork creates a useful screen and provides support for climbing plants.

Left: Polychromatic painted timber adds interest and dramatic colour to this entry gateway.

Taps

One of the most annoying aspects in any garden redevelopment is to end up with water taps in the wrong place. If you leave the plumbers to decide where the taps should go, don't be surprised if they choose the most practical place for quick and easy installation. Think carefully about where *you* want them and make sure they go there.

Bins

Other small details are important too, such as finding suitable areas for storing garbage bins and 'green giant' garden bins. Bins can be an eyesore and should be placed sensibly. The same goes for the compost heap. A composting system is essential for keen gardeners but its location is often a problem. A bin can be a more attractive and efficient alternative to an unsightly heap.

Garden and Tool Sheds

All do-it-yourselfers and enthusiastic gardeners eventually collect more tools and equipment than they can comfortably store. A garden and tool shed not only provides extra storage space but also creates additional work space for a variety of projects, from repotting house plants to carpentry and repair jobs.

Garden and tool sheds come in all shapes and sizes. The design and construction of prefabricated metal garden sheds is a thriving industry. Metal sheds are efficient, highly economical and very practical in design. They can be tucked away in an out-of-the-way corner of your garden and camouflaged by well-designed timber and metal screens.

A simple screen can be created by using two 100 mm CCA (Copper Chrome Arsenic) treated posts set in the ground, with a top and bottom rail of 100 mm × 50 mm timber as a frame for lattice or trellising. With a little encouragement and training, after one season your screen will be covered with greenery and the shed almost completely obscured from view.

Getting the Best from Your Shed

Your garden and tool shed will be doubly useful if you install electric light and power. When you design and lay out your garden, or do home alterations, make sure there is a connection point on the house and a conduit (a rigid tube for carrying and protecting electrical wires or cables) buried in the ground. There is nothing worse than tripping over in the dark in an emergency situation while looking for the right tool!

A basic principle, when designing your shed with extra storage in mind, is to think about the amount of space you will need then double it (if possible)! By the time you have stored the kids' bikes, the lawnmower, a portable barbecue and a few deck chairs, lack of space may be even more of a problem! It is also important to think about your work space needs, especially if you want to install a decent workbench.

The best garden and tool shed will have a waterproof floor. What constitutes the best kind is a matter of opinion. However, if the shed will be used to store wheeled implements such as lawnmowers, access is a critical factor. A concrete slab floor is one answer (see page 91). The precautions you would take when building or installing any additions to your home also apply for a garden and tool shed; for example, see page 24 for commonsense rules on using power tools. Tools and equipment represent a large investment and should be kept in a dry place to prevent rusting.

Above: A garden shed can be camouflaged by ivy and other climbing plants.

Above: Children's toys need to be stored along with garden tools and other equipment.

2

WORKING WITH PROFESSIONALS: A SURVIVAL GUIDE

Guidelines • Council Approval • The Prescribed Payments Scheme (PPS) • Supply and Installation • Insurance and Responsibility • Progress Payments • Surviving Your Building Project

Like all trades and professions the building industry has its own jargon and ways of doing things. To effectively deal with consultants, builders and other trade specialists, it's essential that you are familiar with certain trade practices and terms. There is a comprehensive glossary at the back of the book to assist you and, where special terms are used in the text, they are explained or illustrated.

It's important to know all the pitfalls and how to avoid them. Home improvement projects can be expensive and time-consuming, but they are worth doing, and worth doing well. Whether you do the job yourself or enlist professional help, you need to know what's involved — from getting quotes and estimating and ordering materials to the construction and final assembly stages.

Guidelines

The following guidelines are designed to give you some of the basic information you'll need to plan and organise your home improvement project.

Get It in Writing

In all dealings with the trade you should get a written quotation. The quotation must outline the work expected and the *price agreed* and be accepted, before any work is started. Almost all disputes between builders and their clients can be traced back to a misunderstanding of the nature, scope and cost of services offered and accepted.

Shop Around

Once you have decided on the design and details of your project, ask two or three contractors for quotes — this is called tendering. One of the most difficult parts of tendering these days is finding people willing to give quotes.

With building and landscaping work the best way to find good tradesmen is to ask around. Most people will be willing to express an opinion about their contractor! The best builders don't need to advertise as their work comes from client recommendations and word of mouth.

If you see some work which you particularly like, make contact with the homeowner and ask who designed or built their project. Most people can't resist that form of compliment and are willing to pass on the information.

Obtaining Prices

Make absolutely certain that the contractors who tender for the job know what you want and the standard required. Don't assume that the lowest tender will guarantee the best service and quality of finish. If you are specific about the job requirements you are more likely to get reasonable tenders.

In any tendering situation you will get a range of prices. Consider them all carefully rather than automatically accepting the lowest price. Be particularly wary if the difference between the lowest price and the next couple is extreme. Often the 'bargain' price is the result of a bad mistake or because something has been left out. Like most other goods and services, you get what you pay for, and the lowest price may reflect an equally low standard of work.

Never Pay Deposits

Don't pay deposits to tradesmen prior to their starting work. This is the cause of a lot of problems between small builders and their clients. For a substantial job involving a major order, it may be better to pay the supplier direct for the delivery of materials, otherwise you may find that your money has been used to complete someone else's job!

Whether your project is a deck, a pool or a front fence, careful planning is the key to success.

Council Approval

For any work requiring council approval you will need drawings and specifications which adequately describe the proposed project. Most councils stipulate that any proposed new work is coloured on the plans and clearly marked in relation to the existing structures.

Your plans should be attached to any quote from a builder and the quote should state that the price is based on the plan as submitted. It is often difficult to get a tradesman to quote exactly on the drawings. However, the task of comparing quotations can sometimes seem impossible if too many qualifications are made to the prices.

If your drawings and specifications are accurate and complete the quotes you receive will be more competitive and more comparable. It is wise, in the long run, to employ an architect or draftsman to draw up a plan of your proposed works, for council approval, and as a basis for obtaining prices.

The experienced, professional draftsman or designer will often be able to help solve special design problems and save you money. A tradesman's solution to a problem may not be in your best interests, or even fit in with your design. An accurate plan — which describes what you want built and is the basis of the quoted price — can save a lot of arguments later.

Architects and drafting services will often be able to recommend builders and suppliers as they work in the trade continuously. They have a wide range of trade contacts, and can obtain tenders from a selection of builders. It is an advantage to have a professional overseeing your job, especially if you have problems with the tradesmen. Someone experienced in the trade can usually solve disputes or problems on the job by negotiation.

Your builder may suggest changes to the design and alternative materials as a means of saving you both money. You should be willing to discuss any alternative materials or methods that your contractor suggests. Be aware however that sometimes these alternatives are not in your best interests and won't save money or time. If you are unsure about the pros and cons of a job, ask a professional designer or another tradesman for advice.

Some smaller builders and tradesmen will be happy to let you help and will negotiate their prices to reflect the saved labour. Bricklayers will often reduce their prices if you cart and stack the bricks for them. The footing trenches could be excluded from the concretor's price if you do the heavy work. Plumbers and drainers can also be persuaded to negotiate if you dig the trenches for them. Remember however, that even though a tradesman's rates are quite high, it is sometimes worth paying the extra money. There is no sense in saving money by doing the labouring yourself if you end up paying doctor's bills for treatment of back injuries!

Above: Brick fence and entrance to a courtyard.

Above: A metal safety fence is an important part of pool design.

Above: Painted timber pergola attached to the house.

Some professionals jealously guard trade secrets and will be reluctant to let you help. Others are only too happy to have someone around to be their 'go-fer'. It depends on the individual and the type of job being done. Most of the time you will be able to obtain prices which exclude the heavy labour of digging and carting materials. Your hourly rates will not be all that good, but it is better than paying someone $20 per hour to do it!

The Prescribed Payments Scheme (PPS)

If your project has a total value of over $10 000, new Federal Government regulations require you to submit details of all payments made to builders and contractors to the taxation office.

This legislation is meant to stamp out the previously widespread practice in the home improvements industry of working only 'for cash'. The penalties for non-compliance are draconian, so if your project exceeds that $10 000 minimum you should check with the appropriate tax department office for details of the scheme. The tax office has several useful booklets available which explain the scheme. Your local tax office will be able to answer any queries over the phone.

If you employ a licensed builder for your job and sign a building contract, your responsibilities under the Prescribed Payments Scheme are limited to submitting a form of notification to the tax office. Your builder should be well aware of how the scheme works and will submit his claims for progress payments, accompanied by the required exemption form.

The requirements of the law as it relates to owner-builders are quite complex. If you think that the Prescribed Payments Scheme applies to you (and it will if your next project exceeds $10 000) then you should telephone or write to the tax office for details.

Supply and Installation

It is very often false economy for you to pre-buy materials for a tradesman to install. The professionals can get trade discounts, some of which may be passed on to you. These trade discounts may also involve credit terms. It can cause problems, and even a dispute, when a worker makes a mistake with your materials or if you underestimate the quantity required. For bricks and timber for a larger project and basic materials like sand and cement (usually supplied to the bricklayers and carpenters anyway), this is not so critical. However, to avoid wasting time and money, always ensure that you have a definite arrangement with all workers on your building site as far as responsibility for ordering materials is concerned.

Insurance and Responsibility

As an owner-builder you may be responsible for covering any workers for workers' compensation insurance. If a worker has an accident on your job you could face a claim for hundreds of thousands of dollars. Before you undertake any work or engage tradesmen, find out from your insurance company whether your household policy covers workers' compensation.

If your job requires council approval and exceeds

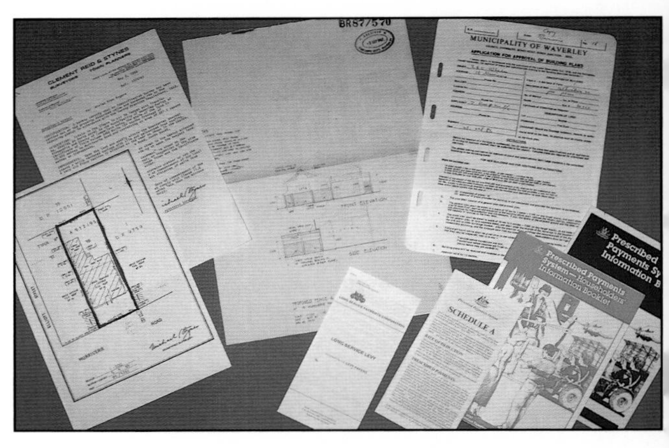

Above: Building projects can involve a lot of paperwork.

Above: Paving for a driveway is a job best done by a professional.

$10 000 in value you may have to contribute to the Building Workers Long Service Scheme. In New South Wales you will need to pay a fee of 0.5 per cent of the estimated construction cost before the council will release the plans! All other states, except Queensland, have similar schemes in operation and you should check with your local council building department for details.

Even if you're not going to employ one solitary tradesman or labourer on your job and intend doing all the work yourself, you will still have to contribute to the scheme. The building industry is used to paying this fee, but many a first-time builder is outraged when his plans are held to ransom by the council (which is obliged to collect the fees).

Above: A successfully completed building project.

Progress Payments

The only really effective way to ensure that contractors complete their work satisfactorily is to ensure that you owe them money!

Progress payments are often misunderstood by both builders and clients alike. The best way to properly assess the value of any work completed is to look at the overall cost to complete the job. At no time should you have paid a contractor more money than the value of work actually completed. If there is 50 per cent of the job to go, the contractor should have received 50 per cent of the money. The only payments made to subcontractors should be for materials actually delivered to site and labour completed. Don't pay for materials or fabricated items 'off-site'. If the subcontractor or supplier goes into liquidation you will have little chance of getting what you have paid for.

Contractors will often ask for progress payments much greater than the value of work completed. This is common business practice and a means of managing cash flow. If a dispute occurs over the quality of work so far completed, the answer from an unscrupulous builder may well be 'We've got your money mate, TOUGH!'. If the builder needs the money it is an incentive for him to finish the job to your satisfaction.

Surviving Your Building Project

The key to any successful building project is good communication. Make sure that the tradesmen understand exactly what you want. Your plans and ideas must be clear and easily understood. If you sketch your vague ideas on the back of an envelope and hand it to the builder to price and build, you will get what you deserve!

Make up Your Mind

Even the most amicable and cooperative builder can turn into an angry, frustrated ogre if clients continually change their minds. The building trade has its own set of management problems. The builder is responsible for bringing a number of people onto a site to do their part of the job. Coordination of electricians, plumbers and a host of other tradespeople is hard enough without constant changes, due to the client's indecision or personal mismanagement.

It is often a fatal mistake to change your mind during construction. The survivor's rule is 'Don't start work until you know exactly what you want'. Altering designs midway will frustrate the tradesmen, disrupt the job's progress and cause a rapid escalation in the cost of the project.

When a tradesman has quoted on a particular job, he assumes that it will steadily progress towards completion. He also has a timetable. Often materials are all ordered at once with the benefit of quantity discounts and lower delivery costs. If materials have to be returned, orders cancelled or additional deliveries made, you will be responsible for these costs.

Clients always complain about the cost of variations to the plan. However, their change of mind can sometimes effectively cancel out the builder's profit for the job. If you choose to remove an item from a contract you may only receive a credit for a fraction of its value. On the other hand, if you *add* the same item to the job it could cost you double!

The best way to ensure that you get good value from your subcontractor is to make a decision and avoid changing the design once construction is under way.

3

WORKING WITH TIMBER

Timber Basics • Ordering and Estimating • Tools • Power Tools • Care of Your Tools • Make it Easy • Ladders • Trestles • Setting Out • Basic Joints • Hardware • Designing Decks and Terraces • Construction Principles • Timber Protection • Project: Building a Deck

The smell of freshly cut timber, curling wood shavings and sawdust is all part of the experience and pleasure of working with timber. The craftsmanship required for building outdoor structures is generally less demanding than for interior carpentry work, but careful work is still necessary for good results. Outdoor timber structures are less formal in their design and 'rough sawn' rather than 'dressed' (or smooth finished) timber is usually adequate. Treated logs and wooden railway sleepers can be used imaginatively if you want something a little different. The larger sizes of timber normally required tend to allow more tolerance for slight inaccuracies or a little inexperience. If you have a good understanding of the basics, a handsaw and a few other tools, you should be able to build an attractive and professional-looking pergola or timber deck. It's all a matter of working slowly and carefully.

Timber Basics

It's important to know the basics about buying, ordering and estimating timber for your project. As with most materials, timber is available from a variety of suppliers. The general rule when buying materials is the bigger the supplier, the better the price. While your local hardware store offers convenience and extended trading hours, it does cost extra. The large timber yards deal directly with timber producers and buy in vast quantities. They are able to offer both range and price advantages. Many owner-builders and complete beginners are reluctant to go to a large builders' yard. Don't be intimidated — the majority of sales people who work at these yards are experienced do-it-yourselfers who will be quite sympathetic to your needs. They are usually willing to give a few tips and select the better pieces of timber for you.

The price of timber and other building materials varies so it pays to shop around. Timber yards should be willing to give telephone quotes. If your order is a big one, post them a list of the materials you need so that they can prepare a proper quote. It's always a good idea to personally choose the actual pieces of timber, especially for a critical section of your job. You will pay a little more for a better grade of timber, but it is really worth it.

Ordering and Estimating

The timber industry uses three basic terms to describe timber quality and a further two to describe the surface finish.

Merchantable: means, for our purposes, average saleable quality; it may have a few knots and not be perfectly straight and true

Select: selected, of special quality and of superior grade

Clear: without knots or blemish; of the highest quality

Sawn (rough sawn): surface finish off the saw, that is, rough sawn texture

Dressed: all four faces planed smooth (smooth finished)

When you order dressed timber you will pay for the full size *before* the timber is planed smooth. Typically, dressing timber removes about 3 mm from each face. A 100 mm × 50 mm sawn dimension becomes about 94 mm × 44 mm after the timber has been dressed (planed smooth). Be careful when you order timber for a critical job as this small difference can cause a big mistake.

The types of timber you decide to use in your garden structures and other work will depend a lot on what is stocked by your local timber yard. Oregon or Douglas fir is one species almost universally stocked throughout Australia. The timber is imported, mostly from the west coast of America, and has become relatively expensive. A local substitute is treated radiata

Above: Treated radiata pine footbridge.

pine with its characteristic light green/grey colour. Local hardwoods are suitable for pergola posts and other structural uses. Visit your local timber yard for a price list and check what kinds of timber are available.

It is sometimes very difficult to correctly estimate the amount of timber and other building materials needed for a project. If you work from a drawing your job will be a lot easier.

Timber is sold in multiples of 300 mm which is approximately equal to 1 foot in the old imperial measure. With 100 mm × 50 mm Douglas fir costing about $4 per metre, you can't afford to overestimate the amount of timber needed for a job.

Professional building estimators make comprehensive lists of the materials required. Some, quite literally, go through a set of plans and colour in each item as it is added to the list.

Above: Milled tallow wood decking provides a very durable surface.

Left: Latticework (treated radiata pine) provides screening, shade and support for plants.

Wastage costs money. It does help to have one extra piece of timber in case you make a mistake, but by adding several shorter bits together you can avoid a lot of waste. As a general principle you should try to order the longest convenient lengths of timber, as well as the lengths that will result in the least total waste. As an example, let's say we have listed the following information for the construction of a pergola:

ITEM	SIZE	FIN	TIMBER	GRADE	NO.	LENGTH
Posts	100 mm × 100 mm	DAR	Oregon	Merch	6 req	2.80 m long

You could order 6 × 3 m lengths for a total of 18 m or as an alternative, order 3 × 5.7 m for a total length of 17.1 m and cut each 5.7 m length in two. You would save the cost of 900 mm of timber (which could be as much as $10) in this order, which would cost around $180. If you repeat this process on all of the different timber sizes to be ordered, the savings could be quite substantial.

As delivery is only rarely free, try to be cost-efficient in your ordering. Most timber yards will charge a single flat delivery fee for each drop. If there are three deliveries in the course of the job, you will be charged three times. Delivery costs up to $15 per drop — if you can arrange for all of the timber to be delivered at the one time, you could save as much as $30. Your local timber yard will also supply a range of accessories. You could make further savings by ordering the footing plates, nails and other small items, along with the timber.

Tools

As you work on various home improvement projects you will gradually build up a good selection of tools. Often a new project will require the purchase of new tools just for that job. Spending money on tools should be regarded as an investment, and good tools, if well cared for, will last a lifetime.

There is an old tradesman's saying which is very true: 'There are only two types of tools, good tools and cheap tools'. Unfortunately, it is very easy to be seduced by the 'bargain' range of tools on sale in hardware stores. Several Australian manufacturers produce high quality tools equal to the world's best. Look at the tools used by the professionals.

As you gain the experience and confidence to tackle bigger jobs you will want a suitable selection of tools. Surprisingly, the difference in price between the 'cheap and nasty' range and the best possible quality is rarely more than 25 per cent. This is a small price to pay for a lifetime's use.

There are only a few tools which are absolutely necessary for general carpentry work around your home and garden. Most homeowners have a set of screwdrivers, a small handsaw and a hammer for those minor repairs. Some tools, like planes and chisels, are fragile and require frequent resharpening, and careful storage. Unless the garden shed is totally waterproof and dry, it will be no place for electrical tools.

A basic carpenter's tool kit will include:

rip saw
cross cut saw
tenon saw
225 mm diameter power saw with carbide-tipped blade
level
tape measure
straight edge and square
10–13 mm capacity power drill
hand drill and set of bits 1 mm through to 13 mm
counterboring bits
brace and bit
mortise and flat chisels 10, 18, 25 mm
mallet
hammer
wood planes (jack plane and smoothing plane)
spanners (for bolts etc.)
trestles and a workbench

Power Tools

Power tools are by no means essential — after all, power tools have only been readily available for less than thirty years! The real wonder is how so much beautiful and sophisticated construction work done in the past, was possible without them. Power tools allow the amateur to tackle jobs that would otherwise

Above: Well-equipped carpenter's tool kit.

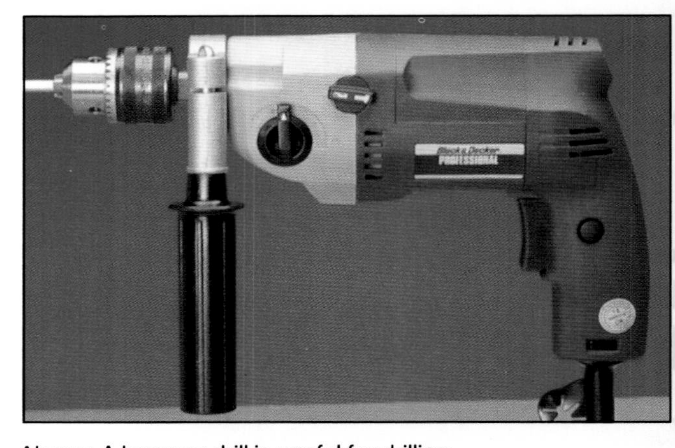

Above: A hammer drill is useful for drilling into masonry.

be exhausting. They also help to achieve better results.

Power tools are potentially dangerous if a few commonsense rules are not carefully followed.

- ☐ All leads should be checked to ensure that they have been correctly installed and that the connections have not become loose
- ☐ Never work while it is raining, or when the grass or ground is wet
- ☐ Wear close-fitting clothing, as a drill can cause serious injuries if the drill bit catches on a loose thread or portion of a garment
- ☐ Always make sure that your footing is stable and that you have enough power cord to reach the place of work

These simple precautions will help prevent accidents.

The first power tool most people buy is a drill. For pergola and deck construction and other basic carpentry work you should buy one of the smaller 'trade quality' models, rather than a light duty type. Drills are rated by 'chuck capacity' which means the largest diameter of drill bit which will fit into the chuck of the drill. The drill motor will also be rated at this size so don't use a larger sized bit with a reduced shank diameter. These bits, although designed to overcome the limitations of the smaller drills, should be

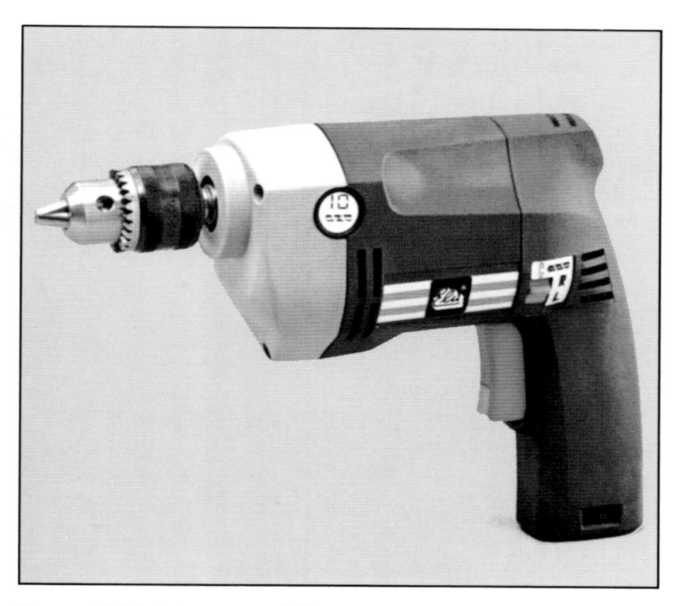

Above: Light-duty power drill.

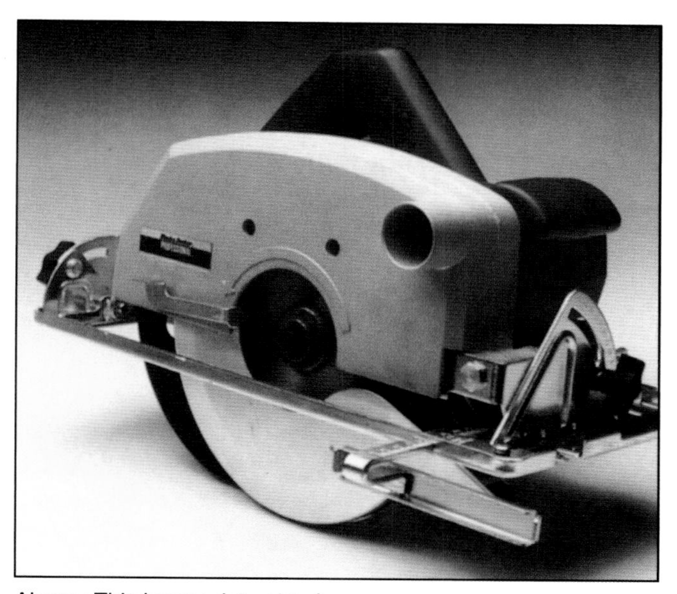

Above: This heavy-duty circular saw features a carbide-tipped blade which is essential for heavy cutting work.

used with caution as the extra load could burn out the drill's motor. If you have to drill a hole with a larger diameter than your drill's capacity, borrow a larger drill or hire one.

The next purchase that a budding do-it-yourselfer might want to make is a power saw. The same rules apply to all power tools. When you decide to buy a tool, buy the best you can afford. It is also important to think ahead to more ambitious future projects. A 225 mm diameter power saw will do for most jobs, but the limit of depth it will cut is about 75 mm. If you eventually plan to erect a pergola, for example, and the posts are 100 mm × 100 mm, the power saw will be too small to do the job safely. The rule should be: when in doubt buy or hire the next size up!

There are many other kinds of useful power tools designed to make your work easier. Whether you buy a new electric tool for a particular job, or hire it for a few days, will depend on the amount of work to be done. If you are going to cut half a dozen housings in a piece of timber for a frame, you can hardly justify the purchase of a router, as the job can easily be done by hand.

One additional power tool which you will find very useful is a bench grinder. This can be used for sharpening all types of hand tools like chisels and plane blades as well as garden shears and knives.

Care of Your Tools

The cost of a decent tool kit can run into hundreds of dollars, so it is important to take good care of your tools. All sharp-bladed tools are fragile and can be dangerous if left lying around. One of your first carpentry projects may be to build a tool box specially designed for your needs.

Chisels are best stored in their own individual compartments. Saws should be kept separate from other tools to protect the teeth. Planes should be stored with their blades retracted, and all other cutting tools should be stored to keep the teeth from rubbing against other tools. For inspiration, take a look at some of the sophisticated tool boxes used by professional carpenters.

If your tool kit is a small one you may be able to find a commercial tool box at your local hardware shop which will do the job. A carpenter's tool box is a very personal thing and its design and workmanship can tell you a lot about the owner. If you have completed one job, and will not be using your tools for some time, it is a good idea to first sharpen your tools and then cover them in a fine coat of oil to prevent rusting. There is nothing worse than starting a new project only to find that all your tools are rusty or blunt. If this is the case you only have yourself to blame. A short spray of WD-40 (protective oil) or a thin film of petroleum jelly will also help to prevent tools from rusting.

The key to maintaining your tools in tiptop condition is to store them sensibly away from dampness and humidity. Also don't force a blunt tool because you can't be bothered climbing down off a ladder to resharpen it. It may seem paradoxical that more accidents are caused by blunt tools than sharp tools — but the extra force needed to make a blunt tool cut, is also more likely to make it slip and cause injuries.

Make It Easy

There will be times when you wish you had an extra pair of hands. For example, hanging off a pergola trying to accurately plumb a post at the same time as attaching the first rafter to steady the whole structure!

Above: Standard extension ladder.

TRESTLE

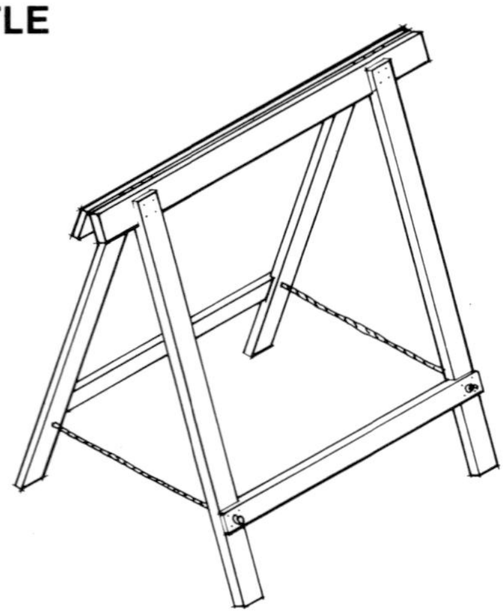

Materials List
75 mm × 50 mm DAR Oregon 2 pieces, each 1800 mm long
50 mm × 25 mm DAR Oregon 4 pieces, each 900 mm long
125 mm gate hinges (galvanised) 2 required
2 dozen 30 mm 8 ga brass wood screws
2 m of sash cord

Tools Required
Tenon saw
28–25 mm wood chisel
Mallet
Wood plane
Hand drill, bits and countersink bits
Screwdriver

The real tricks of the carpenter's trade can often be seen in the techniques used to erect timber structures.

An extra little piece of timber, temporarily nailed in a position to steady a post until it is securely fixed, can make the job a lot easier. Every carpenter needs clamps to act as a second pair of hands. The most common are known as 'G' clamps, and a few 150 mm to 200 mm sized clamps are very useful. The 'quick action' style are excellent in a tight spot and can be applied quickly.

Ladders

Every homeowner needs a ladder, whether for easy access to upper storage cupboards, or for pruning in the garden. Arguments rage as to whether aluminium ladders are better than timber ones. Aluminium ladders have the advantage of lightness. A good ladder will last many years and is well worth the money spent. A large ladder can be a real problem to store, however, you can buy a folding ladder, which is designed to collapse into a convenient storage size. If your storage space is limited, it may be a good idea to look at this style.

For pergolas and other shade structures a lot of the construction work will be high up and a good quality, stable ladder is essential. A sturdy plank placed between the ladder and a convenient resting place, can increase your efficiency many times over. Climbing up and down a ladder all day can be very exhausting!

Trestles

For work around the garden it is useful to have a simple pair of trestles. Two trestles can be made in a morning's work, for a fraction of the cost of buying them. The trestles can double as supports for an outdoor table, and are also strong enough to use as a platform for painting ceilings.

The trestle supports and braces are made from 75 mm × 25 mm DAR Oregon (Douglas fir) and the legs are made from 50 mm × 25 mm DAR timber. You will also need two galvanised hinges per trestle, a dozen brass wood screws and a short length of sash cord.

Connect the tops of the legs to the cross supports with a stopped halving joint. This joint is made by cutting a rebate (a recess, or step, usually of rectangular section) in the supports to half the depth of the timber.

Cut a corresponding rebate from the tops of the legs and screw the hinges through the joint. The screws both hold the members together and attach the hinges. Make the trestle rigid by connecting the cross members between the legs about 200 mm from the bottom. For added support the cross member is then housed over the legs.

All brass screws should be countersunk. This

involves enlarging the upper part of the holes in the timber so that the head of the screw can be sunk below the surface.

These cheap, sturdy trestles will provide a very useful and highly portable work surface for all your carpentry and other handyman jobs.

Setting Out

The way a job is set out reflects the true skills of a builder. Building materials are not cheap, but your own time is. Always take the extra time to measure the timber correctly.

Metric tape measures are often just as difficult to read as the old inch tapes. When you go out to buy a new tape measure make sure it is well designed and readable. (Remember that you will often be hanging on with one hand and trying to measure with the other!) A tape measure which is difficult to read will almost certainly cause errors.

The building industry does not measure in centimetres! For building purposes, centimetres, and more particularly, tape measures graduated in centimetres, are simply an irritation. If you want to buy a new tape measure for your garden project, buy one that is graduated in millimetres and metres ONLY. If you tell the average tradesman that you want something 'X' cm long, the usual result will be confusion. Carpenters work to a tolerance of less than one millimetre. Remember that fractions of units are exactly what the metric system was designed to eliminate.

It is very important to remember how you took a measurement and to cut accordingly. If you measure a length of timber from end to end, the tab of the tape will be placed over one end. On the other hand, to fit a piece of timber into an opening, when you measure, the end of the tape should be inside the face of the opening. Depending on the thickness of the metal end tab of your tape measure, the error could be as much as one millimetre. This incorrect measurement, if not allowed for, will make all the difference between a good fit and a loose one.

Basic Joints

Whether your project is a pergola, a deck or even a timber border, careful work will produce acceptable results. Good tools help, but ultimately it is the way they are used and maintained that will make the difference.

There is no substitute for practice. If you are not confident about completing a joint properly, try it out on a scrap piece of timber first. Very few apprentice carpenters have a good result on their first try, and some, despite plenty of practice, still manage a less than perfect joint sometimes.

The Halving Joint

This joint, as the name suggests, involves cutting a rebate from one or both members, equal to half the thickness. The joint is often used for connecting pergola beams to the posts and in this application has the advantage of temporarily supporting the timber, while the bolts are fixed in position.

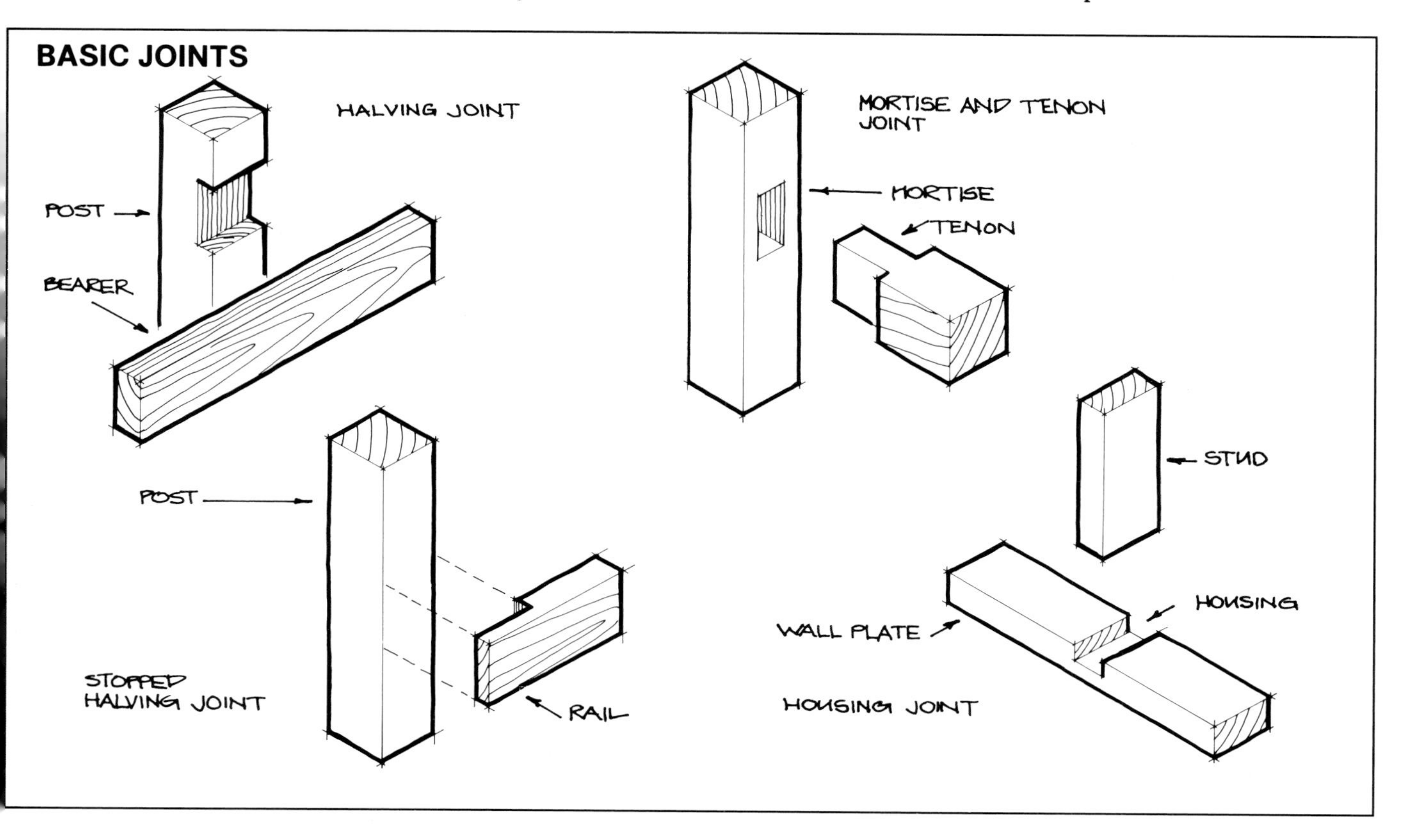

The halving joint is one of the easiest joints to make, and a good carpenter can cut a halving without using any measuring tools. This is because the depth and size of the rebate is determined by the mating piece of timber. Mark out the cuts by carefully positioning the two members together. When you use another piece of timber to mark out a joint, be sure to allow for the thickness of the pencil line when cutting. The general rule is to cut inside the line. Remember, it is always easier to shave a little timber from one member for a perfect fit, instead of attempting to fill an unsightly gap.

The halving joint (especially when it is used for a pergola beam-post connection) needs to be tight and well constructed to add to the overall structural strength of the pergola, and also help to resist wind loads. Good workmanship is always more than aesthetics — a well-made joint is critical to the way the total structure works.

Housing Joints

The housing joint is similar to the halving joint, except that the rebate is generally cut into one member only. Deck rails, like pergola beams, are often housed into the posts.

The advantage of using the housing joint is that only one rebate is needed. This cuts down on work and reduces the possibility of errors. The rebate is cut from the post to assist in the erection of a pergola — this provides a 'ledge' for the beam to rest on while the joint is finished and the bolts installed.

Above: Deck area overlooking the pool.

Mortise and Tenon Joints

A 'mortise' is a rectangular hole cut into timber. A 'tenon' is a tongue-shaped section, cut into the end of a piece of timber, designed to fit tightly into a mortise (see illustration on page 27).

The mortise and tenon joint is most commonly used between the legs and rails of a table. The joint may, however, also be used in pergola construction, as well as in the connection of deck beams and posts. The problem with this joint is that it does not maintain the full strength of the members.

A variation, called a 'haunched mortise and tenon' is stronger, but a little too complex for the average garden structure. The mortise and tenon joint is most often used as a decorative joint, when strength is not the most important consideration.

Hardware

Many of the projects described in this book use metal connectors and bolts to secure joints in timber. The use of these connection systems will make any construction work less complicated. The fittings and plates described in the pergola section particularly, are available through your local hardware supplier. Any special fittings may be fabricated to your design by a local metalworking shop.

All garden structures are exposed to the weather and materials used must have a high level of rust resistance. Any larger metal items should be supplied 'hot dip galvanised'. In this process a coating of zinc is applied to base steel by an electrolytic process. Hot dip galvanising gives superior protection at a reasonable cost.

Solid brass screws are good to use for other minor metal components. Be aware that some fittings sold in sealed plastic bubble packs at hardware shops are only plated steel and not solid brass. The coatings used to mimic brass last only a short time, and will result in rusty stains on your timber which are impossible to remove.

Designing Decks and Terraces

Building a timber deck sounds complicated, but is, in fact, one of the easier home improvement projects. A timber deck provides significant benefits to the homeowner at a reasonable cost. If your home is built on a sloping site, a deck may be your only practical way to create extra outdoor living space.

A timber deck can be attached to the house or free standing. Terracing is also a very effective way of creating additional space on an extremely steep site. The big advantage of terracing is that the construction work involved has only a small impact on the site. With brick and concrete paving, drainage and stormwater runoff can become a serious problem. A timber deck can be erected over the ground level with little or no effect on the surface drainage.

Above: This timber deck provides a delightful outdoor leisure and entertainment area designed to make the most of an ocean view.

Above: Outdoor area with wooden decking.

Below: This pool decking incorporates a platform seat built around a palm tree.

When you build your deck you will be using very similar construction systems as those used in timber houses. The deck will be supported off posts set in footings (usually concrete pads). The posts may be either concreted into the footing pads or bolted to metal plates set into the concrete. Standard metal shoes designed for this purpose are available from hardware and building supply stores. It is better to use the metal plates (or connectors) set into concrete so that the posts finish clear of the ground. This is mainly to protect the timber from termites and wet rot.

At the design stage, think about how your various outdoor entertaining and living areas relate to each other. It is sometimes an excellent space management idea to separate different activity areas by a change of level or area. On a steep site, you may want the barbecue area on one deck and an outdoor dining area close by but on another platform. No matter how large your decking is, there will be occasions when it is overcrowded, for example, when you have a party. The best design solution is to make your deck as large and varied as possible — without compromising other areas in your garden.

Construction Principles

The main parts of a timber deck are the posts, bearers, joists and decking. The posts, bearers and joists are similar to those used in pergola construction. Most deck builders use prefabricated metal connectors cast into concrete pads. As an alternative, you can attach special connectors to a concrete pad by using masonry anchors. It is preferable to cast the fixings into the concrete at the pouring stage, rather than drilling holes in the concrete after it has set.

Framing a deck is almost exactly the same as framing the floor of a timber house. Bearers are connected to the posts in the same way as the bearers under the floor are attached to the brick piers or stumps. Bearers are generally spaced at about 1800 mm to 2400 mm centres. Joists are set perpendicular to the bearers and are spaced at 450 mm or 600 mm centres, depending on the thickness of the decking timber. If your decking timber is cut from 25 mm Australian hardwood, 450 mm spacing is better. For larger spans it is necessary to use decking timber that is at least 30 mm thick. Decking timber is prone to extreme weathering and any deflection or movement of the timber should be avoided.

A deck or terrace which is more than 300 mm off the ground must have a handrail. For maximum safety the railing should be designed with children in mind, for example, to prevent very small children from slipping through the railing. A planter bed made from brick or stone is an attractive alternative to a timber handrail. Handrails have to take considerable strain and the connections to the deck structure must be well designed and put together. Some carpenters use a double joist at each post for additional strength. The advantage of using the double joist method is that the decking can be neatly finished off against the posts.

Details of handrail connection.

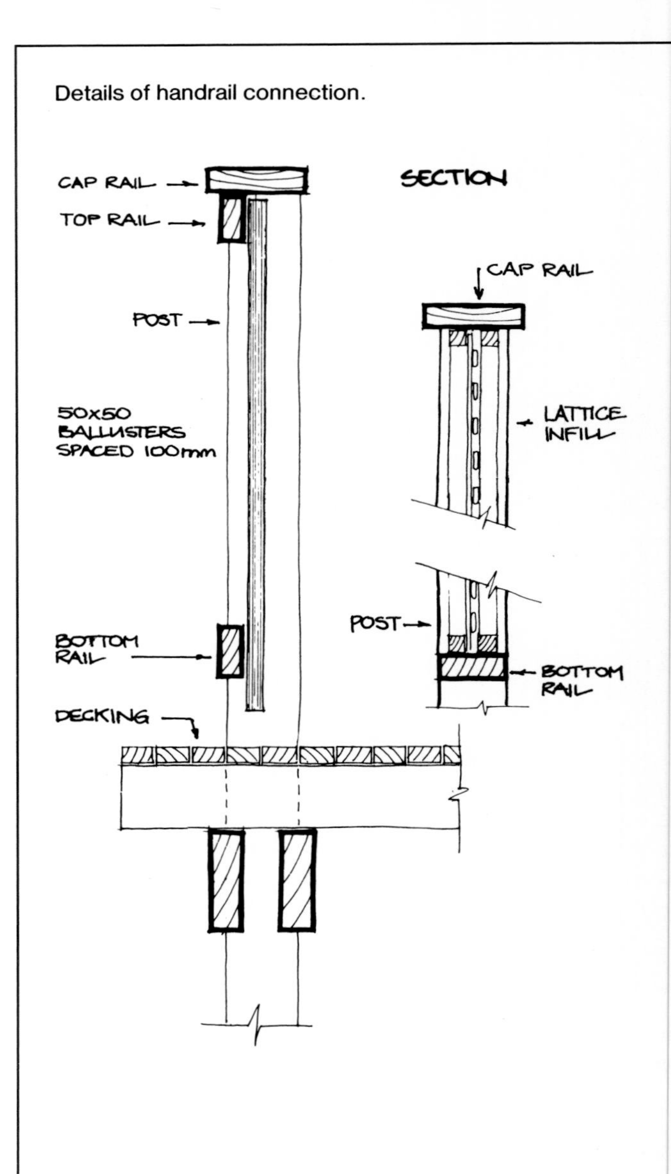

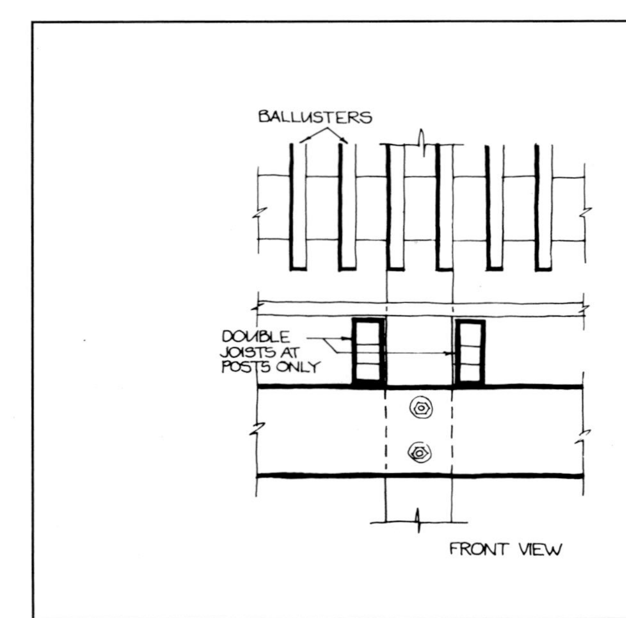

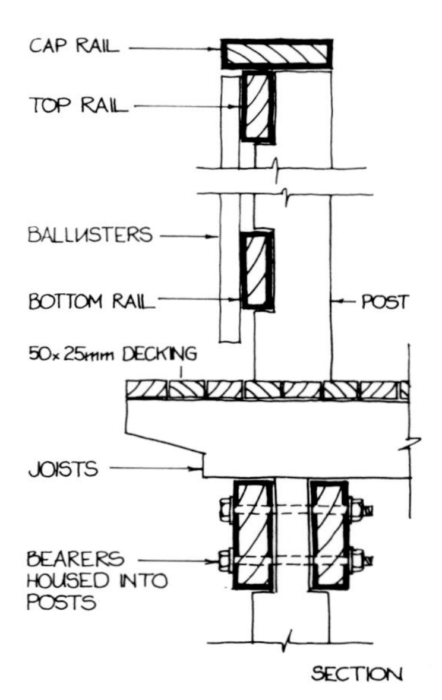

Timber Protection

The best time to stain timber is when it is on the ground prior to construction. Once a pergola or deck has been built it is more difficult to paint. Because of this, the initial coats of protective paint or stain should be applied as thoroughly as possible. Early application of stain or paint will reduce cracking due to rapid drying of timber, particularly hardwood. Remember that when you cut pressure-treated wood the exposed wood in the joints needs to be protected as well. Check the manufacturer's instructions for recommended methods.

As with all outdoor carpentry, the hardware must be adequately treated to prevent rust and corrosion. Use hot dip galvanised bolts and nails wherever possible and ensure that the coatings are not damaged. A common cause of galvanising failure is the use of poorly fitting spanners. If you use the wrong sized spanner it will slip around the bolt head and damage the protective coating. As the base metal is exposed, the process of corrosion is accelerated.

Before you nail down your hardwood decking, drill the holes. This reduces any splitting that may occur, especially if you are nailing close to the edges. Hammer in one nail only at each joist to prevent splitting due to shrinkage. To keep the decking timber evenly spaced use a spacer cut from some thin timber between the boards as you fix them down. All timber will shrink with time so allow a little for this, leaving a space of no more than 2–3 mm.

A deck or terrace doesn't have to be rectangular or square in plan. The less conventional shapes, usually determined by the nature of the site, are often the most successful and interesting.

Terracing in timber decks can be effective as long as it is well thought-out. Decks which are connected by ramps or short stairs on a sloping site can provide very useful and attractive extensions to your living space. Also, because the natural drainage is not affected, timber construction has very little negative environmental impact.

Timber which is buried in the ground must be protected from termite attack. This can be done by using a pressure-treated timber or by treating the soil with the appropriate insecticide. Creosote, an oil-based preservative, can be painted on any timber which will be below ground to provide direct protection.

Any garden structure that you build will not only represent a substantial investment in time and materials but should also increase the overall value of your property. Don't let your well-constructed outdoor structures fall into disrepair through neglect, or lack of proper protection. Because the Australian climate creates a harsh environment for wooden structures timber protection is especially important.

Above: Timber deck and pergola with an acrylic timber paint finish.

Above: This elegant, curved timber deck extends out over the edge of the pool.

Project: BUILDING A TIMBER DECK

The construction of a timber deck follows exactly the same principles as a timber floor inside the home. In a deck, posts take the place of the dwarf piers or stumps.

Posts may be buried in the ground and set in concrete; but the more conventional method is to use post anchors fixed to the footing, which support the posts clear of the ground. This is important in reducing the possibility of insect attack and rotting. The use of post anchors also allows some flexibility in that the posts can be adjusted to a perfect vertical before they are finally fixed into position.

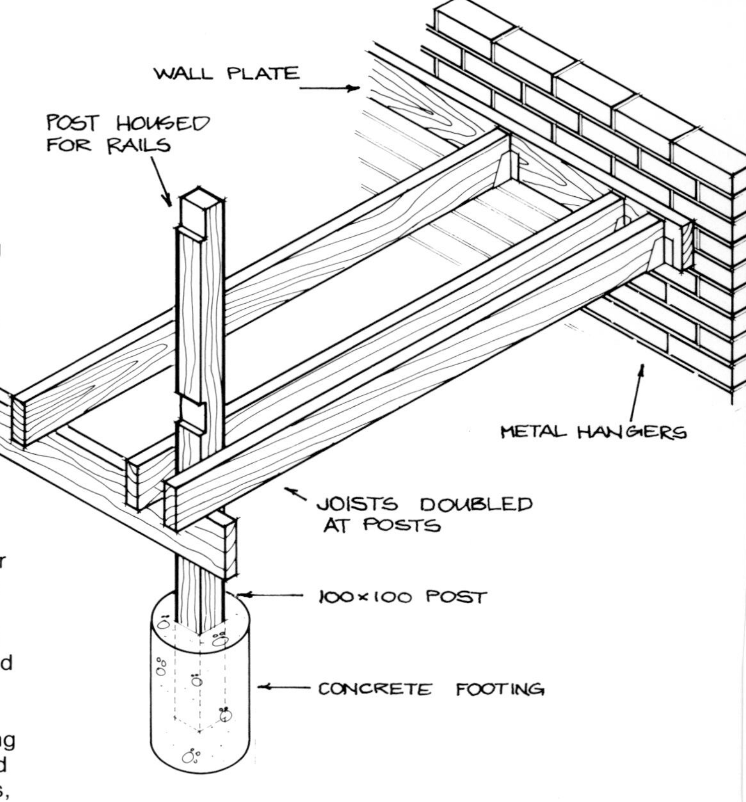

A deck may accommodate many people at any one time, and correct footing design is essential. Depending on whether your deck is to be erected over an existing concrete slab, or whether the construction is all new, several options are available. The simplest footings use a concrete pad. You may choose to build in the steel post connectors, or leave the top of the pad flush with the ground, and drill into the concrete for masonry anchors.

Connecting the deck to the house is a matter of fixing a plate to an external wall. In a timber home, you need to locate the studs behind the cladding. For brick walls, masonry anchors are used. Counterboring the holes for the bolts gives a neater finish, and prevents clothing snagging on boltheads.

The bearers at the outside of the deck are partially housed into the posts and fixed with galvanised steel bolts. Housing bearers and joists in this way make the job of erecting a deck by yourself that much easier, as the timber is self-supporting during building, until the bolts are fixed in position.

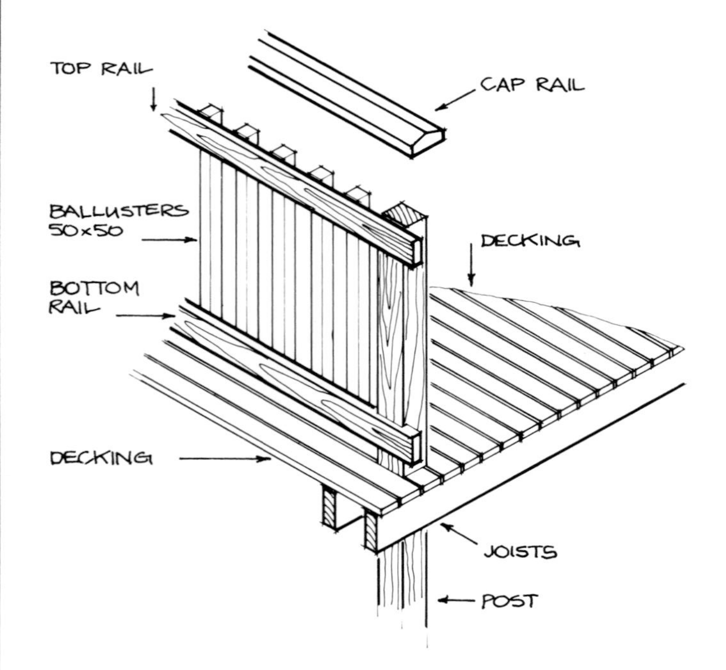

Joists should be spaced no further than 600 mm apart for 38 mm decking, and 450 mm apart for 25 mm decking. Depending on the span, joists will be between 100 mm and 150 mm × 50 mm. Attach the joists to the bearers with two skew nails. Make sure that you use nails which are long enough (100 mm minimum).

Double joists are useful for the connection of the balusters. Connecting balusters to single joists is not ideal as they tend to move.

The major parts of the railing (or balustrade) will be provided by the extended posts. Balusters are made from the same size timber as the posts but do not need to extend to the ground.

Decking is normally cut from Australian hardwood for durability. Nail the decking timber as close together as possible to avoid dangerous gaps.

FINISHING

Generally, decks are made from dressed timber. To reduce the possibility of the timber splitting it's a good idea to chamfer the edges of all members. Stop chamfering results in a particularly neat and professional finish.

Stain or paint finishes are equally suitable for finishing off. The new acrylic timber finishes cover in a single coat. Unusual timber finishes can be achieved with strikingly different colours, so don't feel limited to standard brown!

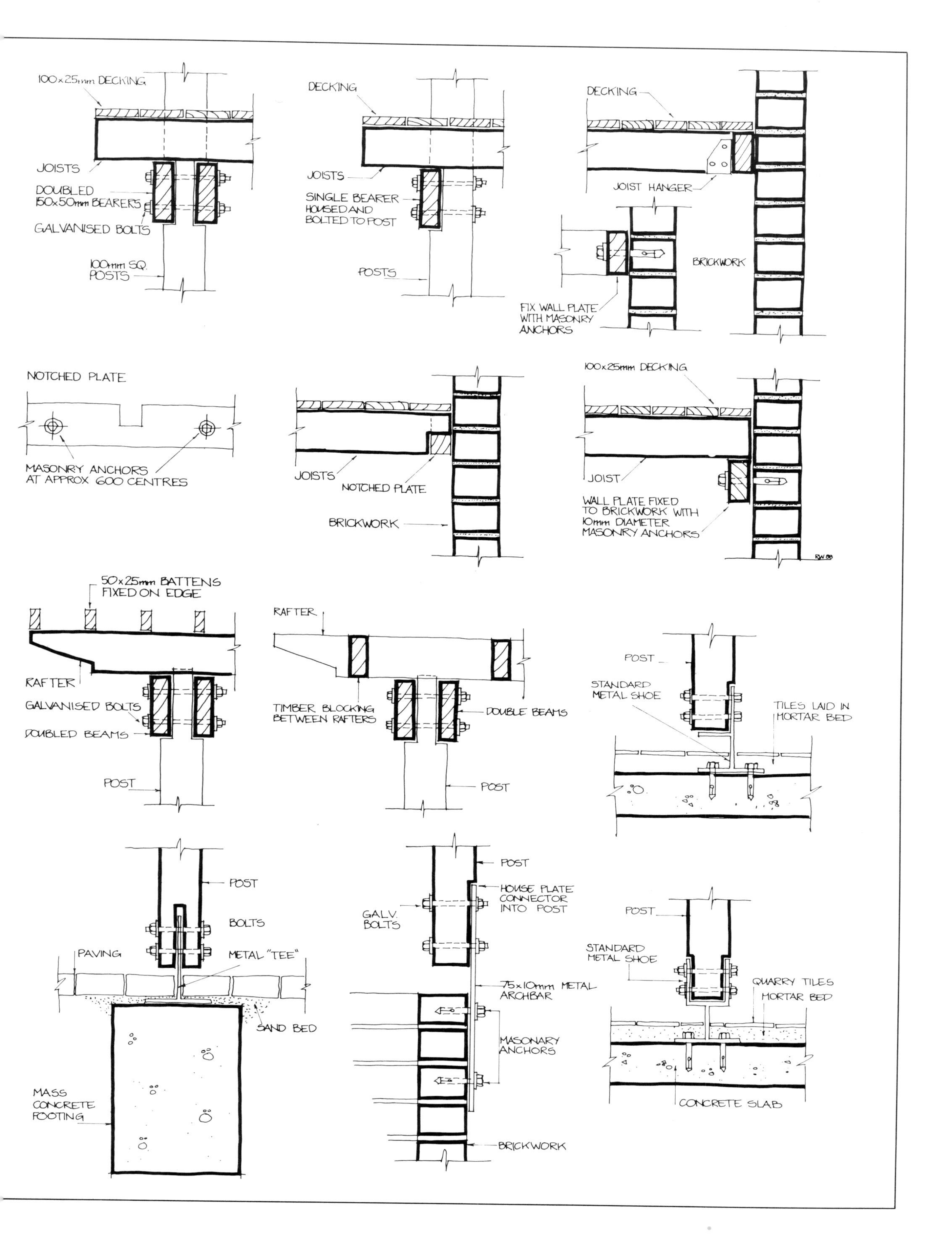
100x25mm DECKING
JOISTS
DOUBLED
150x50mm BEARERS
GALVANISED BOLTS
100mm SQ.
POSTS
DECKING
JOISTS
SINGLE BEARER
HOUSED AND
BOLTED TO POST
POSTS
DECKING
JOIST HANGER
BRICKWORK
FIX WALL PLATE
WITH MASONRY
ANCHORS
NOTCHED PLATE
MASONRY ANCHORS
AT APPROX 600 CENTRES
JOISTS
NOTCHED PLATE
BRICKWORK
100x25mm DECKING
JOIST
WALL PLATE FIXED
TO BRICKWORK WITH
10mm DIAMETER
MASONRY ANCHORS
RW 88
50x25mm BATTENS
FIXED ON EDGE
RAFTER
GALVANISED BOLTS
DOUBLED BEAMS
POST
RAFTER
TIMBER BLOCKING
BETWEEN RAFTERS
DOUBLE BEAMS
POST
POST
STANDARD
METAL SHOE
TILES LAID IN
MORTAR BED
POST
BOLTS
PAVING
METAL "TEE"
SAND BED
MASS
CONCRETE
FOOTING
POST
HOUSE PLATE
CONNECTOR
INTO POST
GALV.
BOLTS
75x10mm METAL
ARCHBAR
MASONARY
ANCHORS
BRICKWORK
POST
STANDARD
METAL SHOE
QUARRY TILES
MORTAR BED
CONCRETE SLAB

4

FOOTINGS AND RETAINING WALLS

Types of Walls • Basics of Footing Design • Concrete Footings • Brick Footings • Levelling the Footing Trenches • Digging the Footing Trenches • Reinforcement • Retaining Walls • Materials for Retaining Walls • Project: Building a Retaining Wall

A footing is the lowest part of any structure — the part that transfers the total load of the structure to the ground — and is usually made from reinforced concrete or bricks. If your foundation is rock you can build a brick wall directly on the rock after first scabbling it (rough levelling). The terms 'foundation' and 'footing' are often used interchangeably. However, 'foundation' correctly refers to the part of the ground which bears the load of the footing, and then the whole structure.

The suggested footing designs shown in this book are for footings most commonly used in construction work. Every site is different, however, and it is important that you consult with a structural engineer before applying any of these techniques.

Your local council will be able to recommend several professional engineers in the area. Also try the state office of the Institute of Engineers and the *Yellow Pages*.

A retaining wall is a wall specially designed to hold back earth or soil. Retaining walls, even small ones, often have to support quite heavy loads and structural quality is a critical factor. Common materials used for building retaining walls are stone, brick and concrete blocks and second-hand railway sleepers.

Types of Walls

Walls can be divided into three categories, all of which need a footing: low walls, medium to high walls and retaining walls. A low wall, with a height of less than 1 m, won't generally need the structural sophistication required for a medium to high wall (between 1.5 m and 3 m). In most cases your local council will insist that the footing design for a fairly high wall be checked by a structural engineer. This applies particularly to retaining walls which have to support substantial loads.

Above: Reproduction cast iron infill to a concrete block fence.

Above: Rock faced, coursed stonework.

Above: Retaining wall to a planter featuring recycled bricks.

Left: Fieldstone wall.

Basics of Footing Design

Each different type of foundation has the capacity t support a certain maximum load which is known a the safe bearing capacity. Solid rock has the highes bearing capacity and swampy ground has the lowest.

Some soils are reactive. This type of soil swell when it's wet and shrinks when it's dry, which raise special problems for the engineer. Soils which have a large organic content are also difficult design problems. Your house would have been virtually impossible to build successfully if there were problems with the soil — however if you're unsure, seek advice from an engineer.

In the first instance, look at other walls in your neighbourhood. If your next door neighbour has a wall about 1500 mm high your land should be able to support a similar wall. The maximum safe bearing capacity of the ground in your area will determine the size of the footing needed. A 1500 mm high brick wall made from double brick construction will weigh about 1 tonne per lineal metre. If the safe bearing capacity of the foundation is 2 tonnes per square metre then you will need a footing some 500 mm wide. Allow a little extra for the weight of the footing itself — the suggested footing width would be 600 mm (about 2 ft).

If possible, don't build your wall too close to a large tree. If you do build close to a tree you will have to cut back the roots to stop them from undermining the footing and eventually causing damage. You will also have to increase the amount of concrete and steel for the footing. Tree roots are strong enough to move and ruin even the sturdiest fence or wall.

Concrete Footings

Concrete footings act as beams, and the depth of a footing depends on the wall. For a simple wall a footing depth of 300–450 mm should be sufficient. A concrete footing should be reinforced with steel. For most simple and medium to high walls (less than 1500 mm) standard trench mesh will do. Trench mesh is a prefabricated steel wire cage specially designed for use in footings.

Ready-mixed concrete is relatively cheap to buy, especially if your job requires more than a couple of cubic metres. Depending on the distance the truck has to travel from the yard to your site, the cost of ready-mixed concrete will be around $120 per cubic metre delivered. Carefully setting out and excavation of the footing trenches is important because it can mean a significant reduction of concrete needed for the job. When you are going to cut a trench for the footing, make sure that you have some spare timber to line the trench as formwork. If your soils are self-

Left: Lattice panel in a painted common brick wall.

Below: This front wall features painted, bagged brickwork.

supporting the sides of the trench will stay vertical until the concrete is poured, and there will be very little wastage.

It is sensible to order the concrete on the same day as you complete the trench. A lot of first-time concretors forget to allow for over-excavation of the trenches when they pre-order the concrete. If you carefully measure the completed trench you can accurately calculate the real amount of concrete you need (rather than the theoretical amount). Allow for at least 10–15 per cent wastage.

If you haven't calculated the amount accurately and need extra concrete, you'll have to make a frantic call to your supplier. If your supplier can't give you the small amount needed to complete the job you may have to make a construction joint. (A construction joint is not really a good alternative in a footing as the beam should act as one unit.)

A construction joint is a position, in a normally continuous concrete pour, where it is convenient to break the pouring operation. In concrete paving this is achieved by placing a section of edge board along a predetermined line, to finish the concrete. When the concrete has cured, a few days later, the pouring can resume. Positioning of construction joints must be carefully thought about and not done on an ad hoc basis.

Brick Footings

Many simple structures like planters and small barbecues don't need elaborate reinforced concrete footings and for low walls a simple brick footing will often do.

As a general rule, for a wall made out of single or double thickness brickwork, a footing constructed from the same bricks as the wall is more economical (and you won't have to wait a week for the concrete to set before starting the brickwork).

In any brick wall built off a brick footing, your progress is limited only by the number of bricks you can lay in one day. Also there is the practical limit of not being able to lay more than about 10 courses per day, so that the wet mortar at the bottom is not stressed or squeezed out by the weight of the brickwork above.

Levelling the Footing Trenches

The accuracy of your completed brick wall will depend on how accurately you have levelled the footing trenches. On a sloping site it is necessary to 'step' the footing in amounts equal to brick courses. For example, if a wall 5 m long is to be built in a location where the ground slopes 600 mm along the full length, the footing would have seven 86 mm steps equally spaced from end to end. There would actually be six separate levels in this footing. Alternatively, you could make some of the steps as two courses rather than one. The 86 mm dimension is one standard brick plus the standard 10 mm mortar joint.

Above: Low brick retaining wall laid in English bond.

The reason for building the footing in steps equal to a 'brick dimension' is to avoid cutting bricks later. Any minor inaccuracies can be taken up in the thickness of the first mortar laid directly on the levelled ground and in the first couple of courses. The important thing is to ensure that the first course of the wall above the ground is 'dead level' and to work carefully from there, to maintain that level.

Rather than using a surveyor's level (called a 'dumpy level' in the trade), tradesmen use a 'water level' to check whether long dimensions are dead horizontal. Water always finds a level. A water level uses this principle. It is simply a long length of clear plastic tube filled with water. The tube should be long enough to extend the whole length of the longest level check that you will need to do. Few jobs would require more than 10 m of tube. If the tube is carefully filled with water so that about 150 mm of air is left at each end and there are no bubbles, an accurate level check will be possible every time.

For the simplest low brick wall laid from a single skin of bricks the footing would be quite satisfactory if laid up in one course of 'brick on edge headers'. A header is a brick laid perpendicular to the other bricks, or across the line of the wall. A brick 'laid on edge' is laid on its face. The face of a conventional brick is 76 mm high by 230 mm long. The footing described above would be 110 mm high (the width of the standard brick) by 230 mm wide. The 100 mm thick standard brick wall would be erected on the

Above: Substantial front brick wall with recessed planter.

centre line of the footing. This simple footing is suitable for brick walls up to about 1,200 mm high.

For taller walls, or those built from double brickwork, a more substantial footing is necessary. To keep the loadings on the ground the same as for the simple wall described above, the footing would be made twice as wide. In practice however, the 230 mm wide footing is over-designed. The normal design for a double brick wall of no more than 1,800 mm in height could be safely erected on a 350 mm wide by 230 mm deep brick footing.

A 350 mm × 230 mm brick footing is constructed in two courses of bricks. The 350 mm dimension is one full brick plus one half brick, including the 10 mm mortar joint. As before, the bricks are laid as headers, on edge rather than 'on flat'. To make the footing stronger, the vertical joints should not be continuous. This is achieved in two ways. The second course is laid so that the vertical joints are over the middle of the bricks in the course below. In addition, full bricks are laid over the half bricks of the lower section.

For retaining walls and walls which are high, complicated or have to be erected on suspect foundations, most local councils will insist on the footings being designed by a qualified structural engineer (look in the *Yellow Pages* under 'Structural Engineers').

Above: The niche in this rough, bagged brick wall incorporates a security grille and supports a pergola.

Digging the Footing Trenches

Footing trenches must be dug to a depth below the existing organic matter in the ground and to an undisturbed section of the foundation. In most gardens the depth of topsoil won't exceed about 600 mm (2 ft). Old roots and other vegetable matter will die and eventually rot, leaving voids under the footing. The footing may not work very well if this kind of material isn't removed.

If the topsoil and subsoil are self-supporting you may not have to use edge boards or formwork. In dry, sandy soils however, the sides of the trench will have to be formed up. Second-hand galvanised corrugated iron is often used by bricklayers and concretors to form the sides of their footing trenches. The iron is cut into strips about 600 mm long, and driven into the ground. Backfilling behind the iron will give sufficient support while the wet concrete sets. A few cross braces will stop the iron from moving inwards while the concrete is being poured. This is a rough but simple and economical method which gives excellent results.

The bottom of the trench is the part of the foundation which will support the whole load. You must try not to disturb this at all. The sand and soil that will inevitably fall into the trench should be stabilised by ramming the earth with the head of a rake. A special ramming tool can be made from a sturdy handle and a strong, flat piece of timber.

Reinforcement

A qualified structural engineer should design the footings for any wall over 1500 mm high. The engineer's design will specify the type of reinforcement steel needed. The normal type of reinforcement used for a simple wall is trench mesh (a prefabricated steel wire cage specially designed for use in footings). Trench mesh is available in several grades, according to the dimensions of the footing, and the load the footing is designed for. Trench mesh is available from the larger hardware stores and specialist reinforcement suppliers such as Humes ARC.

Your design should also specify the amount of 'concrete cover' needed. Concrete cover is the thickness of concrete that protects the steel from rusting. This is an important design specification and cover for footings should be not less than 50 mm (2 in). Specially designed concrete reinforcement 'chairs' are used to raise the steel the correct distance from the ground. The chairs are formed from steel wire or plastic and are placed beneath the steel at regular intervals. Small pans, which look like paint tin lids support the sharp ends of the chairs. The pans and chairs are available from most building supply stores Be sure to specify the *amount* of concrete cover required when you order.

Retaining Walls

Terracing is an effective way of increasing useful garden space. Most terracing projects will require the construction of a retaining wall. Broadly speaking a retaining wall may be defined as a wall which holds back earth or soil. Simple, low retaining walls made from brick, stone, concrete or timber are easy to build, bearing in mind that the work is heavy and demanding. If your retaining wall is to be higher than 1200 mm (4 ft), the design should be checked by a qualified structural engineer.

The loads which even a small retaining wall has to support are quite considerable and the consequences of a mistake could be severe. The greatest test of a wall's structural quality is when there has been a lot of rain. To reduce the effect of ground water, designers use agricultural drainage lines (piece of pipe, either plastic or rigid UPVC, with slots punched into it; or flexible like a concertina) behind the wall as well as generous weep holes to allow excess water to drain off.

Retaining walls are normally built with a 'batter'. This is when a wall has a slope built into it. A properly designed retaining wall will have a slight slope backwards against the ground to help resist overturning.

Below: Treated log and stonework retaining walls are used here to create a system of terraces.

Above: Rough, weathered stone retaining walls.

Materials for Retaining Walls

The most common material used for building retaining walls is stone. This may be used in a variety of forms. The cheapest (and also the most time-consuming to lay) is random-sized stone without cement in the joints. These dry-jointed walls have stood the test of time, although the technique is mostly used for low walls. The use of mortar as a jointing material gives added strength.

Stone can be ordered with two sides 'faced' or 'squared', or may be cut to exact dimensions. In common with brickwork, stone layers lay their walls in courses (continuous, usually horizontal, layers). Stone walls may be 'random coursed', which means that the stones have been squared and cut to dimension, but the lengths of the individual stones are not equal. Dimension stone is more expensive than random cut stone, and the general rule is that the harder the stone and the more cutting involved, the greater the cost.

A larger wall will be, in most cases, built by a professional stonemason. Look in the *Yellow Pages* for a local stonemason. If you see some good workmanship try to find out who did it, and where they can be contacted.

Brick and concrete blocks are also common retaining wall materials. Concrete footings are commonly used, although for smaller walls bricks and blocks alone can be used. In all brick wall construction, drainage and weep holes should be kept clear of cement mortar and the agricultural lines checked for effectiveness.

As a general guide, a brick wall less than 1200 mm high may be built off a 350 mm brick footing for two courses, and the remainder completed in 230 mm (double brick) work. As the height increases, the thickness of the base of the wall must be increased. For example, a 3600 mm (12 ft) high brick retaining wall may be built off a 600 mm × 600 mm reinforced concrete footing, and the first 1200 mm section will be four bricks (470 mm) thick. The next section will

Above: Treated timber logs have been used as the main landscaping feature on this steeply sloping site.

Below: Log retaining wall with handrails designed to extend the deck area.

be three bricks thick (350 mm) and the final section will be in 230 mm brickwork. Remember that a standard metric modular brick is 230 mm long, 110 mm wide and 76 mm high. A substantial wall of this nature will use about 650 bricks per lineal metre, not including wastage!

To reduce the amount of expensive brickwork in a larger retaining wall, designers often use reinforced concrete blockwork faced with bricks. Consult your architect or engineer for details on these types of walls.

Precast concrete interlocking 'crib' walls are also popular as plants can be put in the wall's recesses and will eventually obscure the face of the wall.

Second-hand timber railway sleepers are a useful material for the construction of retaining walls and the rough weathered look is popular with landscapers. They are often available free from local railway yards. New timber sleepers are also available from specialist landscapers.

In retaining walls sleepers are usually interconnected with steel pipe driven through holes drilled into the timber. In higher walls, 'whalers' are used to stabilise the loads. These are timber sections inserted at right angles to the wall and buried deep behind it. For additional strength, a pair of whalers may be connected together. A recent innovation has been the use of CCA (Copper Chrome Arsenic) treated pine in a specially designed retaining wall system.

Above: Sleeper retaining wall.
Below: Building a log retaining wall.

Project: BUILDING A RETAINING WALL

Building a low brick or timber retaining wall is a comparatively simple project. Higher walls should be designed by professionals as the forces in a substantial wall are considerable.

The ground must be excavated down to well-consolidated earth or rock if available. The foundation must be firmly rammed to compact it. The ground at the rear of the wall needs to be excavated back at the natural angle of repose so that it doesn't collapse into the trench during construction.

Two or three courses of bricks are used for the footing, each successive course reducing in thickness by one half brick, until the dimension of the final wall is reached.

Every second perpend (or vertical joint) in each second course of the wall must be left free of mortar to assist drainage of water through the wall. The wall may have header courses in each third course to help tie the two leaves of brickwork together.

The wall is finished off with a final course of brick on edge headers (bricks laid on edge rather than on the flat). This gives a neater finish to the wall than simply finishing off with a final course of stretcher bond. Modern bricks often are cored, rather than having conventional 'frogs' in one face, and so are more difficult to finish off neatly.

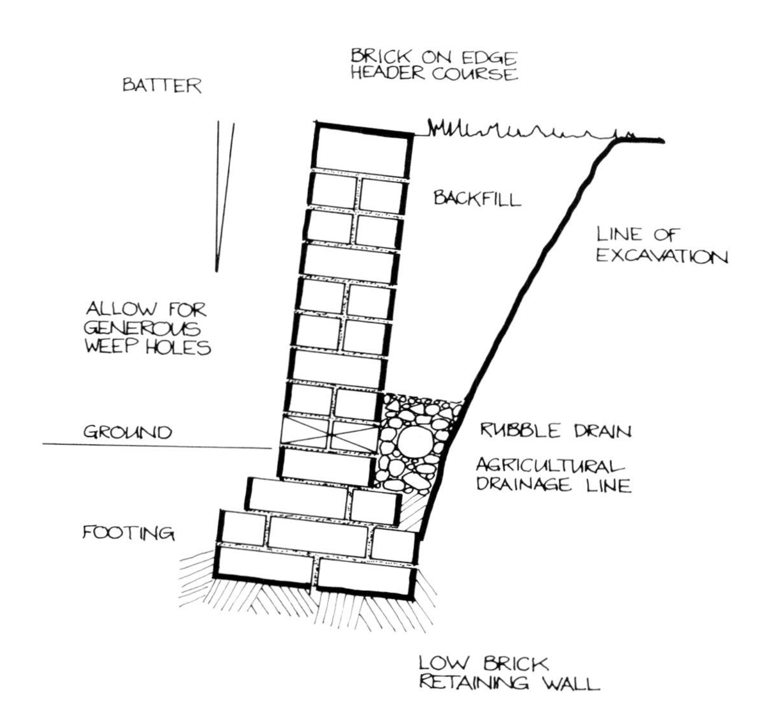

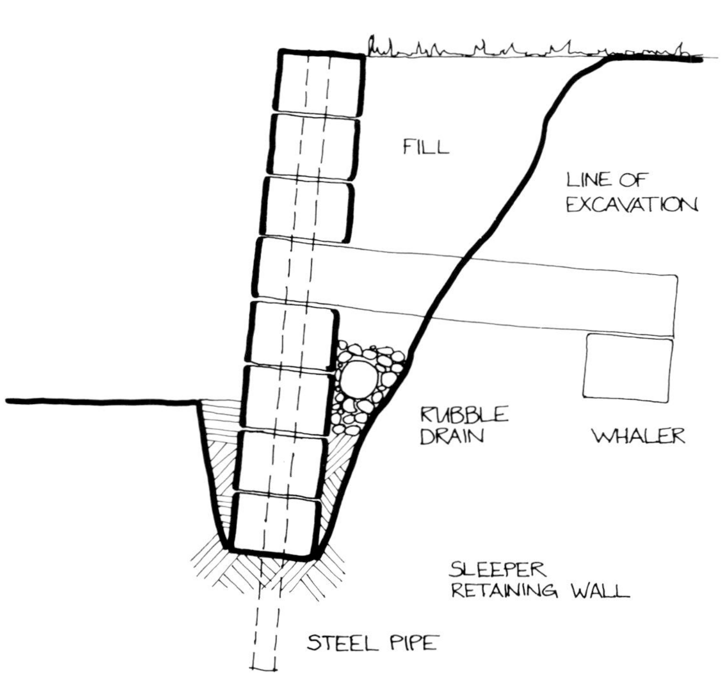

SLEEPER RETAINING WALLS

An alternative detail is to use new or second-hand railway sleepers as the retaining wall material.

Excavation and site preparation are the same, but the timber may be fixed in position with steel pipes fitted through pre-drilled holes and spiked through into the ground.

Additional strength is provided by whalers , which are fixed at right angles to the wall at intervals of about 2400 mm. The whalers are placed deeper into the earth behind the wall and provide further resistance to displacement.

Timber sleepers may be obtained from nursery suppliers, or direct from the railways.

5

WORKING WITH BRICKS

Brick and Mortar Basics • Types of Bricks • Choosing Bricks • Second-hand Bricks • Estimating the Number of Bricks You Need • Buying Bricks • Delivery • Mortar • Mixing Mortar • Coloured Mortars • Project: Building a Brick Planter

Most keen do-it-yourselfers, and especially landscape gardeners, will at some time need to lay a few bricks. Bricks are attractive, economical and easy to handle. Their surface is little affected by weather and they come in a wide range of different colours from white to dark brown. Bricklaying is satisfying, hard work and it pays to know a few of the basics before you start.

Brick and Mortar Basics

Bricks and mortar are the basic materials used. Mortar is mixed from cement and sand, and sometimes a little lime is also added to alter the characteristics of the mix. Bricks and mortar are both heavy materials and bricklaying is very hard labour because it involves substantial handling of these materials.

One of the basic rules with bricks is that you shouldn't mix different styles and sizes of bricks because of the difficulties of maintaining 'bond' — which is the arrangement of bricks in a wall in which they overlap so as to provide structural strength. Bricks should be laid in a pattern so that the vertical joints in any one course don't coincide with the bricks in the course directly below — if they did it would create structural weaknesses in the wall. If you have a supply of bricks of varying dimensions, use them as part of the footing.

Types of Bricks

Bricks are made from clay, concrete, calcium silicate and even mud! They are all made by subjecting the raw materials to heat, and sometimes pressure, to form the brick into its final shape.

For clay bricks, the greater the heat and pressure applied the harder the brick becomes. Over-burnt bricks are called 'clinkers' because they make a clinking sound when knocked together. Clinkers are often so hard that it's virtually impossible to drill holes in them.

Under-burnt bricks are softer than normal bricks and clinker bricks. Although this makes them easier to work with, they are also inherently weaker and tend to crumble.

Concrete bricks and Calsil (calcium silicate) bricks are more consistent in their make-up. Calsil bricks are made from sand and lime, and are white in colour.

Above: Clay brick paving courtyard.

Above: Painted brick storage bin with timber lid extending out from the wall.

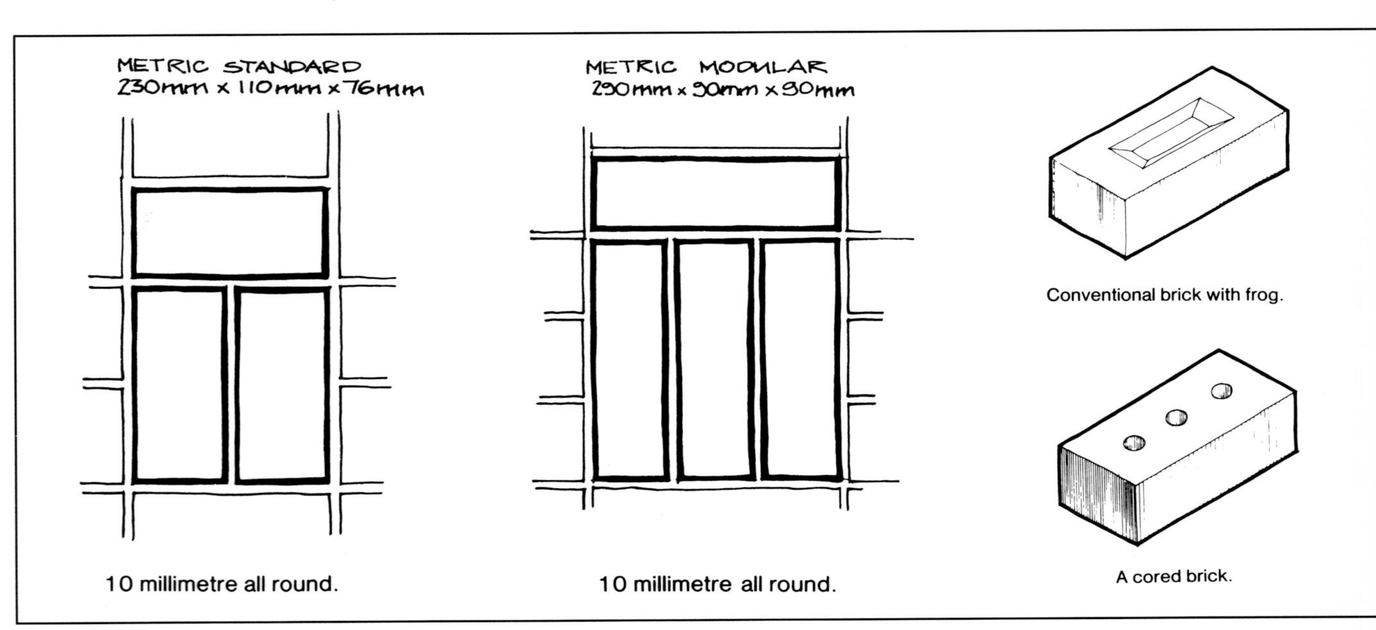

10 millimetre all round.

10 millimetre all round.

Conventional brick with frog.

A cored brick.

Common building bricks are cheaper and are generally used where appearance is not important. They are ideal for footings and where the final finish is paint or render.

Face bricks are designed with appearance in mind. There is a multitude of colours, textures and finishes available. A basic rule to remember when choosing face bricks for a project is that for small areas of face brickwork a contrast is acceptable, but for larger areas you should always try to match the rest of the house.

Other types of bricks include fire bricks and engineering bricks. Fire bricks may be necessary for the hearth of a larger barbecue, although well-burnt face bricks or clinkers will normally do the job. Engineering bricks are tough and waterproof and are generally used underground or in structural applications where exceptional strength is needed.

There are two major sizes of brick available on the market: the 'metric standard' brick and the 'metric modular' brick. The differences between the two are important.

The metric standard brick is 230 mm long, 110 mm wide and 76 mm thick — the metric equivalent of the old '9 in' brick, which was 9 in × 4½ in × 3 in. The metric modular brick, introduced at the time of metric conversion, is 290 mm long, 90 mm wide and 90 mm thick.

A standard joint of cement mortar is 10 mm thick (a little less than the old ⅜–½ in thick bed). The brick dimensions are designed so that the width of two half bricks, plus the 10 mm joint, will equal one full brick.

Above: Stained timber pergola built off a face brickwork dwarf wall.

Below: Clay paving bricks and timber latticework combine to make a very pleasant little courtyard.

Choosing Bricks

The type and style of bricks you choose for bricklaying projects will depend on your own taste and the type of work to be done. For brickwork that will be painted or rendered, the best choice may well be the cheapest acceptable brick available — so shop around. Metric standard common or selected face bricks can be used for most projects around the home.

Bricks for face brickwork need to be carefully selected. Anyone who has built a house will know that choosing bricks can be quite bewildering because of the variety of styles available. If you are planning a substantial structure, such as a new front wall, look at a wide range of completed brick walls before you make a decision.

Choosing the suitable style of bricks is very much like choosing paint colours. One sample brick can look very different when multiplied and made up into a wall! Ask your brick supplier to give you some addresses where the particular style has been used. If you go to an exhibition site or a building centre display take special note of the jointing techniques used, and the colour of the mortar. All of these things affect the final appearance of a finished brick wall.

Second-hand Bricks

Once of the most difficult tasks you may have as a home renovator is matching existing brickwork. If your home is over 10 to 15 years old, chances are that the bricks used are no longer made and you will need to look to the second-hand market.

There are many demolition yards that specialise in second-hand bricks and may have what you're looking for. Some companies also cater for the renovation market, both for the general sale of second-hand bricks, and the matching of rare or unusual bricks. Second-hand bricks usually cost less, add character to your home if used imaginatively (even when cleaning doesn't remove all the paint and mortar from their previous life), and are especially good for paving.

Above: Convict bricks.

Above: Second-hand bricks.

Below: Overburnt clinker bricks.

Estimating the Number of Bricks You Need

It's surprising how many bricks it takes to make even a simple brick structure. In a small or more complex job, allow more for breakage and cutting of bricks. For example, the more corners your structure has, the more you will be cutting and fiddling around to get a perfect fit.

In a straight section of wall, 'laid up' as a single skin of brickwork, about 45 metric standard bricks are needed for each square metre. For an average job, allow for 10–15 per cent wastage. If your wall will need 10 000 bricks, allow for less than 10 per cent wastage.

Face brickwork is quite simple to estimate, but the footings are more difficult. Footings are the part of a wall that is below the surface of the ground. The footing can be made from two or three courses of brickwork laid as headers (bricks placed across the line of the wall) — it depends on the design of your wall. Sometimes a concrete strip footing will be used.

The easiest way to estimate the number of bricks and the amount of other materials needed for any job is to draw a diagram of the work showing a typical 1 metre length. Calculate the number of bricks needed for this section and multiply by the overall length of the wall, to get the quantity you need.

Every time you cut a brick there will be some wastage, and often the bit left over is unusable. Bricks are sold in pallet lots, or more commonly, in thousands. The price varies greatly from location to location, and delivery will be charged for. It makes good sense to shop around.

Buying Bricks

For many people, the job of choosing bricks for a major project is difficult, whether it's a new front fence or just a simple barbecue. There are literally hundreds of brick manufacturers and thousands of styles, textures, colours and other features to choose from. It will help if you can compile a short list of the styles you like.

In the end it's a question of personal taste, combined with what you think will suit your style of house, even your lifestyle. Some houses may be spoilt by the addition of red texture brick details. Alternatively, the fashionable sandstock bricks, so popular in inner city suburbs where there is a real mixture of old and new architectural styles, may look all wrong when used with a very modern style in a new suburb. Make sure you look at a wide range of completed brick walls and their settings, before you make a final decision.

Delivery charges for bricks, as with most other building materials, are quite expensive. Organise your job so that all the bricks are delivered in a single truck

Above: Splitting concrete pavers for a driveway and carport project.

load. For a larger project requiring many thousands of bricks, deliveries may be staggered so that the job site is not too congested. If you're going to use a special style of brick, any variations in colour may cause problems. Negotiate with your brick supplier to stockpile enough bricks from one batch, to avoid matching problems.

Delivery

Large quantities of bricks are usually delivered palletised by specially equipped trucks. This method saves the supplier time and money (compared to the old method of hand unloading and stacking) and also benefits you because the quantity delivered will be the same as ordered, and there will be a lot less damage.

It's always a good idea to shop around for your building materials. When you get quotes be sure to ask for a cost breakdown so that you know what is included in the price.

With the advent of palletisation brick manufacturers have become more stringent about minimum order quantities. The extra cost of broken pallet lots can be very expensive. There are specialist suppliers, and the larger builders yards keep small quantities available for the handyman and home improvement market. Buying a brick individually can cost more than $1 compared to only 20c if you buy in thousands. For a project using less than 100 bricks the convenience of buying them at the local hardware shop could be worth the extra cost. For a larger project however, the purchase of one pallet at least is recommended. A few hundred bricks left over from one project can be used to complete a couple of smaller ones.

One solid dry-press brick weighs about 2–3 kg (about 4½–6½ lbs)! If you want to transport bricks in your car, be very careful not to overload it. The average family car can only carry about 50–100 bricks and even then you must take care to distribute the load evenly.

A wheelbarrow is essential for moving bricks, bags of cement and sand around on site. If you don't have a wheelbarrow you can hire one at a very reasonable cost from a builders hire yard. If you want to buy one, for projects and general use around the garden, go for a sturdy, trade quality model. Wheelbarrows get a lot of rough treatment on building sites, and light-weight models have a short life!

Below: Face brickwork planter finished off with a capping of special paving bricks.

Above: A sturdy wheelbarrow makes the heavy work of paving easier.

Mortar

Mortar is the binding agent for sticking the bricks together. It's a mixture of sand, cement, lime and water, plus some other additives, mixed in defined proportions depending on the job. Mortar is informally known by bricklayers as mud!

The consistency, texture and wetness of the mortar have to be carefully controlled for the best results. Complicated chemical reactions take place as the mortar changes from a wet mixture to a rock-like material.

The strength and flexibility of the mortar is dependent on the amount of water added to the mix and, more significantly, the proportion of cement and sand. For most bricklaying purposes there are two basic types of mortar mixes used by bricklayers: cement mortar and 'compo' mortar. Cement mortar, made from sand and cement, is the stronger of the two and is used for most walls exceeding 1500 mm in height, retaining walls, and in footings below ground level. For other types of work, such as barbecues, low planters and walls, it is more economical to use 'compo' mortar which is made from cement, lime and sand mixed in water.

Mixing Mortar

To make cement mortar, sand and cement are mixed in the proportions of 4:1 by volume. Compo mortar is mixed from sand, lime and cement in the proportions of 6:1:1 by volume. The use of lime makes the mix more economical, but gives a weaker and more flexible mortar which is suitable for smaller jobs.

A gauge box is a very useful tool for measuring out

correct proportions for mixing. A clean dry bucket will do the job as long as you don't use the same bucket for the water. Only add enough water to make the mixture workable. Many first-time bricklayers make the mistake of mixing the mortar too wet. This makes it hard to get even joints, and the excess water stains the bricks.

If you don't use enough water it will also cause problems as the mortar will be too stiff to work, and will give a weak joint. Properly mixed mortar has a dampish clay-like smell that you will learn to recognise.

It's difficult to specify a correct amount of water to add to the mix. The amount varies according to the wetness of the sand and the quality of surface you mix the mortar on. Temperature and humidity are also factors. With a little practice though, you can accurately estimate the correct amounts.

For small jobs, mix the mortar on the ground. You can use a section of concrete path although this will be difficult to clean afterwards. A square section of waterproof plywood, old formwork plywood, or a section of clean steel sheet are also good, easy-to-clean surfaces to work on.

Use a bucket or a gauge box to measure out the quantities of sand and cement then thoroughly mix them together DRY first. If the sand has any lumps break them up with the back of your shovel. Keep on turning it over until it is completely mixed. Next, make a hole in the middle of your pile and empty about half a bucketful of water into it. Mix in the water, adding a little at a time, until the mix is sufficiently wet to be workable. The ideal bricklaying mix is a little sloppy, but never wet. The mortar should be quite stiff.

Only mix enough mortar for your immediate needs. If you can't use all the mixture in one hour (less in very warm weather), then throw the batch away and mix a fresh one. Don't retemper the mix by adding a little more water as this will weaken the mortar too much.

The work of mixing reasonable quantities of mortar on the ground is backbreaking work. However, for small jobs you won't use a lot of mortar — especially if the work is complicated, for example, a small barbecue or detailed work in a front fence project.

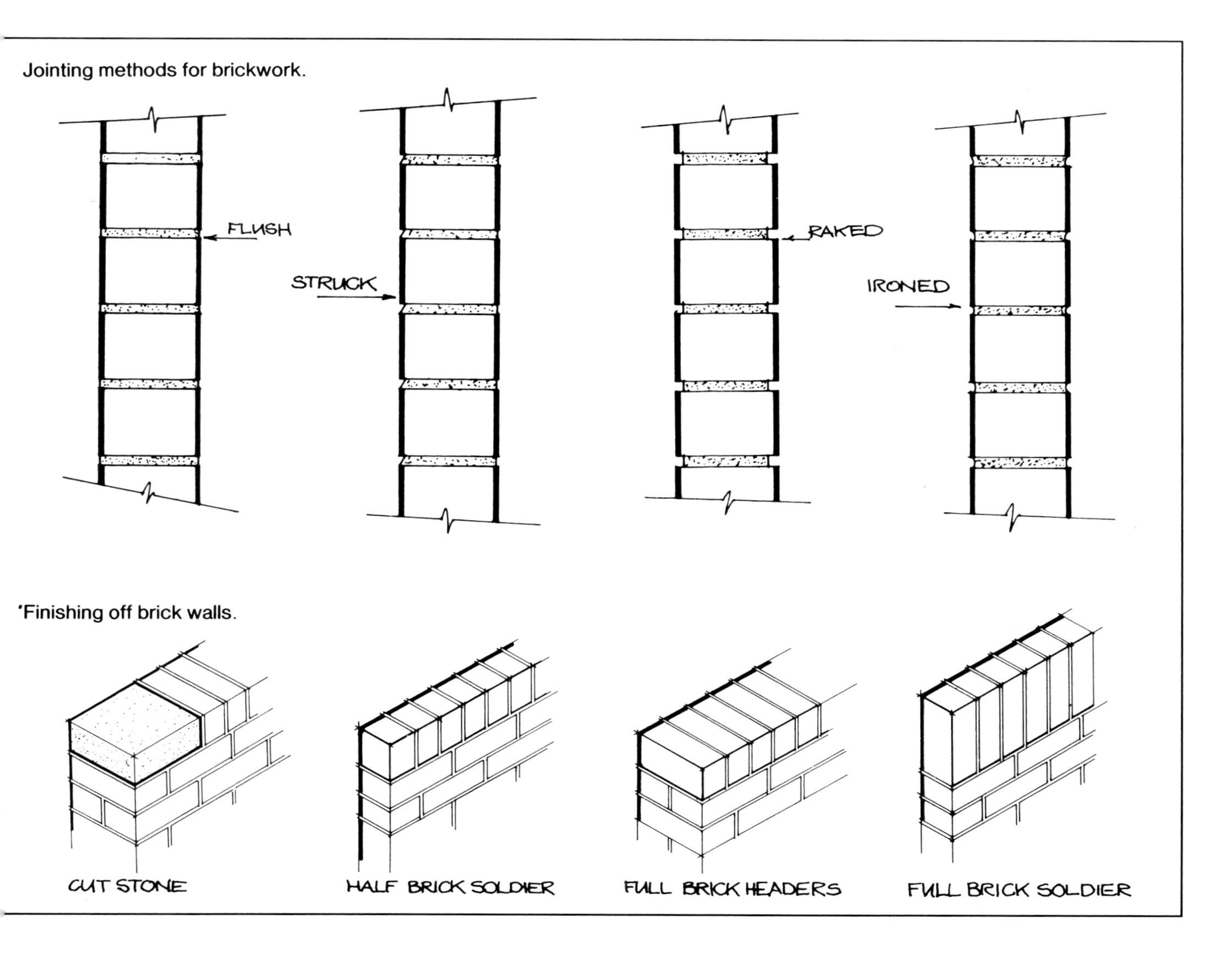

Jointing methods for brickwork.

Finishing off brick walls.

Mixing Mortar in a Wheelbarrow

Professional bricklayers mix mortar for small jobs in a wheelbarrow. This has several immediate advantages. Double handling is eliminated as the mortar may be taken straight to the job site rather than being shovelled from the ground into the barrow first. Cleaning up is also easier as the barrow is simply hosed out afterwards.

Mixing mortar in a wheelbarrow with a shovel is not easy work. Bricklayers use a special tool called a 'larry' which is the secret of quick and easy mortar mixing. A larry looks like a long-handled hoe, with several holes in the blade. It is used in a 'to and fro' motion, and as the wet mortar extrudes through the holes in the blade it mixes thoroughly.

Even if you only use a couple of wheelbarrow loads of mortar for your project, a petrol or electric driven cement mixer will really help. In a power mixer you can be sure that the mixing is thorough. Also, by filling the mixer with the buckets of sand, cement and lime (if any) it is more likely that the proportions will be correct.

Coloured Mortars

For some work you may need a coloured mortar to match existing walls in your home, or to create special effects. Special pigments designed to colour mortar and cement grouts are available from most hardware stores. You may need to experiment before starting the job so that the colour of the mortar, when it dries, is a good match.

The manufacturers of these products have useful hints and guides and you should follow their instructions closely. Before you decide on the proper mix proportions, and the amount of colour to add to your mix, allow the test batches to dry out thoroughly for comparison. When you're satisfied that the colour matches, take care that your mix remains consistent once the job has started.

Project: BUILDING A BRICK PLANTER

Planters come in all shapes and sizes. This simple design is for a free-standing brick planter with one rounded end. A single brick on edge footing is used.

The footing trenches should be excavated down to undisturbed ground, and all loose material removed. The footing is laid in a cement mortar (four parts sand to one part cement). The base course should be as level as possible as the accuracy here will affect the rest of the work.

The brickwork can be done with any type of brick, and it can be left as face brickwork, or may be bagged or rendered.

Weep holes are essential for the health of the plants to be planted in the planter. All vertical joints in the brickwork must be left clear of mortar for the bottom courses and especially the first course out of the ground.

Broken bricks and other clean rubble is needed to fill the bottom of the cavity to further assist drainage. The depth of the soil and therefore the depth of the planter itself will be determined by the size of the plants to be used. The larger the plants, or the bigger the planter, the deeper the soil will need to be.

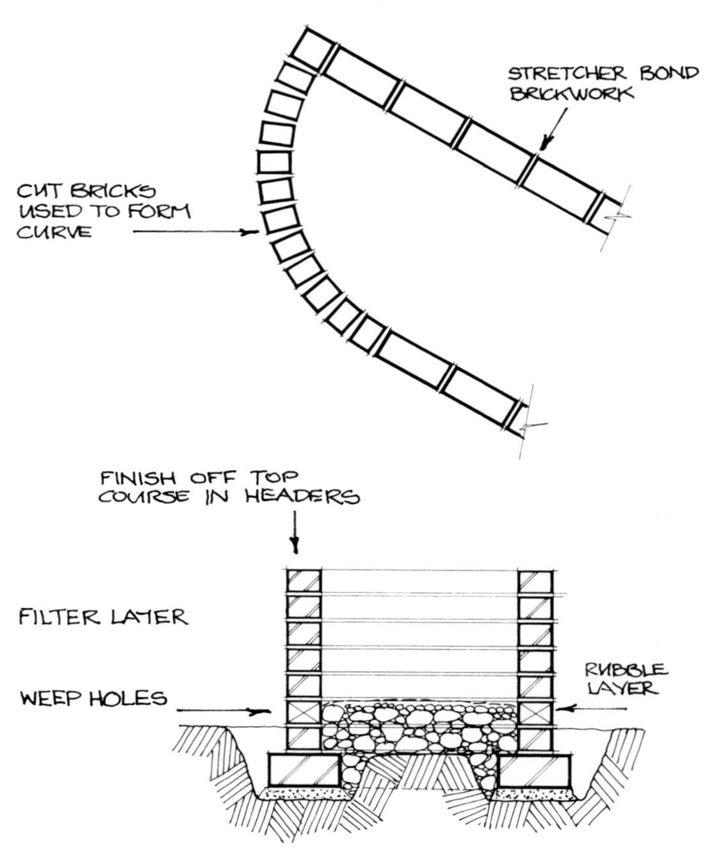

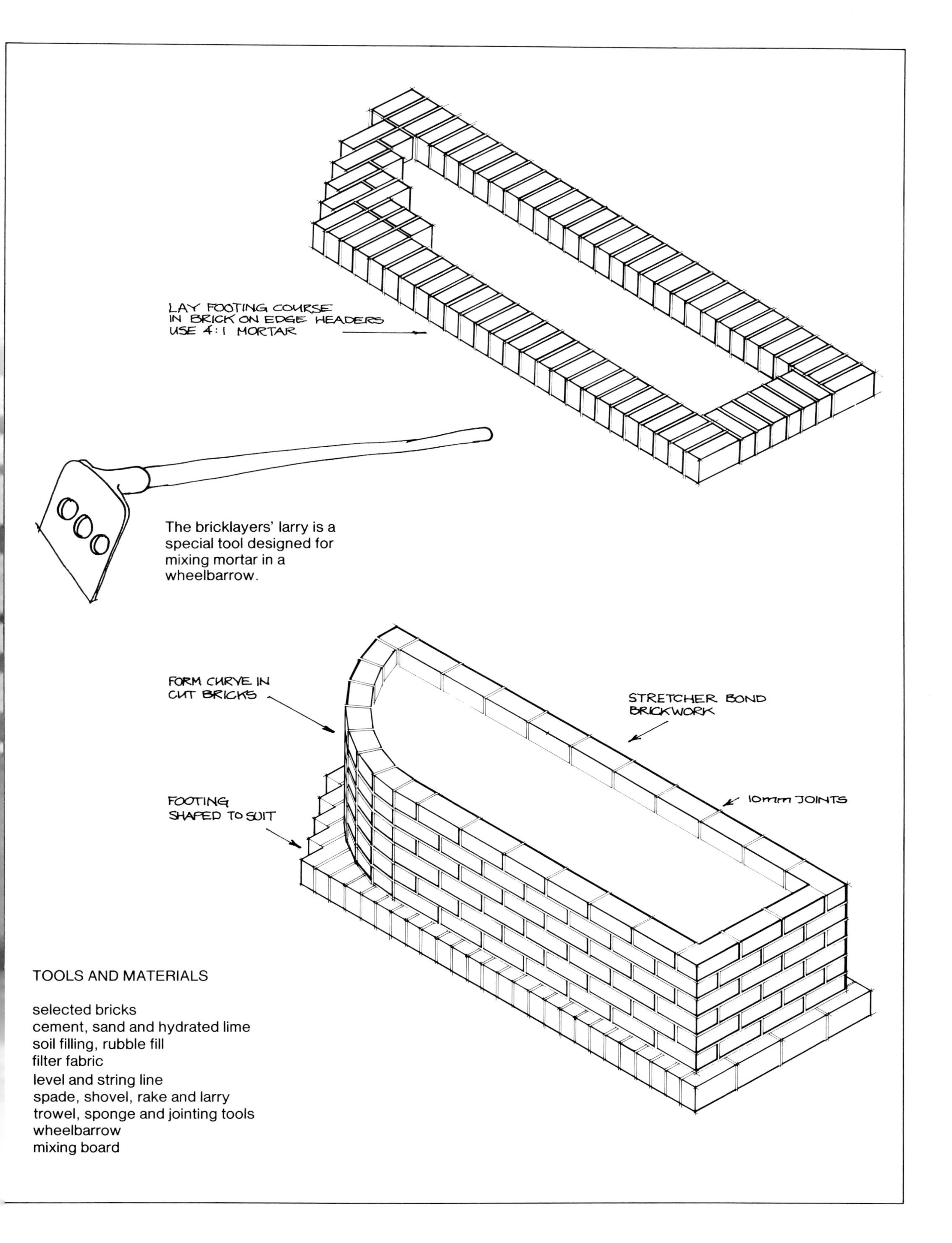

TOOLS AND MATERIALS

selected bricks
cement, sand and hydrated lime
soil filling, rubble fill
filter fabric
level and string line
spade, shovel, rake and larry
trowel, sponge and jointing tools
wheelbarrow
mixing board

6

FENCING

Fence Design • Composite Design • Council Approval • Brick Fences • Timber Fences • Design Ideas • Fence Styles • Tea-Tree Fences • Metal Fences • Pool Fences • Gates • Letterboxes and House Numbers • Guidelines for Building a Timber Fence • Project: Building a Simple Timber Fence

A well-designed fence, especially a front fence, is an important part of the overall design and 'look' of your home — whether it's mainly for privacy, security, sound isolation, marking a boundary or simply to keep out the neighbourhood dogs! No matter what your needs are, this chapter provides some basic design and construction guidelines for you to follow.

New fashions in fence design have resulted in an exciting range of fences to choose from. The old timber paling style of fence will always be popular. Lattice, trellis, wrought iron and cast aluminium, as well as steel and tea-tree infills, are all materials available in a variety of different fencing styles. Your choice of fence will reflect your individual taste as well as the style of your home. Your fence doesn't have to 'match' your house so much as harmonise with all your other design features.

Above: Driveway entrance framed by a pergola and a picket fence.

Above: Brush fencing.

Below: Brick and timber lattice used together make a light and secure fence.

Fence Design

The first thing to think about when designing a fence is its purpose. What are your priorities? Privacy? Security? A wind, sound, sand or light barrier? If you only want to mark the boundary lines of your property a simple, low fence will suit your needs. For inner city residents security has become a serious concern, and high, substantial brick and composite material fences are being built for greater security and privacy.

A high front fence will not necessarily deter a burglar. Once an intruder has climbed the wall and is inside your garden, the wall helps rather than hinders him, because he is not visible from the street. A 1.5 m wall will give the same degree of privacy and protection as a 1.8 m wall and the intruder will still be seen from the street.

Fence height is an important urban design consideration. The increasing number of high fences and walls has made our streets less attractive and more impersonal. Local councils are trying to modify fence designs with community interests in mind.

The overwhelming visual impact of a high front fence can be reduced in several ways. Other shapes can be incorporated into the wall, for example, recesses for plants which will eventually obscure the wall. Other materials, such as tea-tree infills, can be combined with an otherwise full brick structure. Combining materials and varying the texture and form of a front fence will make it more interesting and improve the look of your home.

Even if a front fence has been designed with your privacy in mind, it should welcome, rather than intimidate your visitors. If the entrance way is hard to find, or if your visitors have to wait out in pouring rain while the remote-control latch is operated, your home will seem unwelcoming. If you want to install that level of security, consider how the system will affect both welcome and unwelcome guests. If you do have a security system think about some form of covered entry way — an entry portico can be a delightful addition to a well-designed front fence.

Composite Design

Many fence designs now combine materials like bricks and timber lattice to create a look of delicacy and lightness. Timber lattice is very popular now, probably as a result of the current appreciation of Federation-style architecture. Modern timber protection methods, such as CCA (Copper Chrome Arsenic) treatment, provide excellent weather resistance and also stop termite and insect attack. Lattice is often inserted into timber frames fixed inside brick piers.

Timber lattice and trellis will eventually weather to a light grey colour if left untreated, or may be stained or painted. Tea-tree and melaleuca fencing is also being used to break up monotonous surfaces and to harmonise with less formal landscape designs. These two materials are normally installed by specialist contractors and you should seek professional advice about installation.

Another popular fence design, inspired by oriental garden architecture, consists of narrow slats of stained or painted timber fixed to horizontal rails, and set very close to each other. The slats are usually 50 mm × 25 mm pieces of dressed western red cedar placed about 15–20 mm apart. This design is well suited to smaller gardens. Although it is quite expensive in terms of both materials and labour, the delicate and interesting effects are well worth the cost.

Council Approval

Building regulations are different according to the state or area that you live in. Always visit your local council and discuss their requirements for boundary fences. Generally, if you only want to replace an existing fence, you won't need to get council approval — however, always check first.

The popularity of high boundary fences for security and privacy has caused local authorities to stiffen regulations governing front fence construction. Sometimes approval of a proposed front fence won't be automatic. Even if your neighbours have high fences, don't assume that council will let you have the same style of fence. Some inner-city residents have become concerned at the number of high walls in their areas — to the point of pressuring local councils to stop the trend.

Some councils will place a height limit on a brick wall for the front boundary. Others require a more varied fence or wall design, combined with landscaping details, to create a less imposing, more attractive effect. Urban design is very much a council responsibility, and most councils have quite comprehensive policies for fence construction. A discussion with your local building inspector, or a visit to the council engineer's department, will give you a better idea of what's allowed in your area.

If you want to replace the fence between your property and your neighbours', you may need to make a building application. The first, and most diplomatic step, is to consult your neighbours, and come to an agreement. The Boundary Fences Act in New South Wales, and similar legislation in other states, has formalised a system for resolving disputes between neighbours. If you share a fence with several homeowners, it is wise policy to proceed with tact and understanding. Conflict, especially if the matter ends up in court, can be very unpleasant, time-consuming and costly. In the case of repairing a dilapidated old fence, the law sets out methods by which you can recover the restoration costs. Find out about the fine legal points from your local council.

Above: Lattice infill can vary the look of a solid brick wall.

Above: Front brick wall incorporating an attractive recessed planter.

Incidentally, it's very foolish not to refer a proposal to council for approval. Firstly, penalties for erecting illegal structures are severe. The council can demolish the structures and recover their expenses from you! Secondly, if you decide to sell your house, any prospective buyer will want the assurance that your house and all existing structures on the property comply with council regulations. If the council building inspector visits your home and finds any illegal structures, it could jeopardise your sale. The council can also insist that you demolish the structures. To save all kinds of problems later, make an application to council for permission to build the fence. The same advice is valid for all garden structures of a substantial nature.

Brick Fences

Building a new, brick front fence is a fairly substantial undertaking. Let's assume that your house has a frontage of 20 m (65 ft). Your fence will be made from double brickwork 1.8 m high (6 ft), and allowing for a concrete footing, a gate and the driveway, will need about 9000 bricks. The concrete footing will be about 600 mm × 600 mm, and will need about 5 cubic metres of ready-mixed concrete.

An amateur bricklayer working on his own would be hard-pressed to lay 500 bricks over a very long and tiring day. If you have to cart the bricks, mix the mortar and lay the bricks, the job could take you at least a week of heavy, strenuous work. It's worthwhile getting a quote from a team of bricklayers. For smaller jobs, where the work is quite straightforward, the rates should be around $300–$500 per thousand bricks. Naturally these rates are negotiable and will depend on site access and the nature of the work. If you're prepared to get all the materials, and to act as a labourer on site, you may save some money. Digging the trenches for the concrete footings and pouring the concrete yourself, will also reduce the cost.

What Sort of Footing?

As a very general rule, if your fence is less than 1200 mm in height, it can be laid on a brick footing (see Chapter 4). If you want to build a relatively high fence, over 1500 mm in height, a reinforced concrete footing would be better. Some local councils will insist that you submit plans prepared by a structural engineer, as well as a formal building application. Check your local council's requirements.

Building a concrete footing is not too difficult and can be cheaper and easier than a brick footing. Buy the appropriate reinforcing mesh used to stiffen the footing, from your local supplier. Ask for trench mesh and specify the size of the footing when you order it. For most brick fences, a footing of approximately 300–450 mm deep by 450–600 mm wide, will be adequate. Get your ideas checked and drawn up by a structural engineer.

Timber Fences

Timber fences can range from the simplest paling style (which is almost universal in Australia) to the more elegant variations on the traditional lattice and trellis styles being designed today. Whatever the design, there are a few basic principles which, if adhered to, will help you to design and build a strong and long lasting timber fence.

The Australian climate is particularly harsh on external timber work. Combined with some unique insect pests, such as termites, climatic factors can destroy certain types of timber at an alarming rate. Fortunately, recent developments in timber treatment have almost eliminated these types of problems. With sensible design, and proper selection of timber types and protection, a new timber fence will last well. Some manufacturers are offering guarantees, for up to twenty years, against rotting and termite damage.

Below: Painted timber picket fence with graduated colonial style palings.

Left: Basket-weave interleaved timber fence.

Below: Bamboo is an interesting and effective fencing material.

The Timber Development Association has put out a number of useful publications — about the proper design and care of timber structures — which can be obtained from your state branch or local building information centres.

Design Ideas

A timber fence is a simple structure made up of vertical posts, horizontal rails and some kind of infill. The infill may be traditional timber palings, trellis or lattice, diagonal slats, timber, bamboo, tea-tree, fibre-cement products, metal mesh or a host of other materials.

One way to find ideas for your new fence is to look at what your neighbours are doing. Keep your eyes open, note the good ideas, and just as importantly, the very bad ones. A good fence will be a feature in itself. It must suit your style of architecture and coexist with your neighbours'. Even if a particular material has featured on the front page of every house design magazine for the past few months it won't necessarily suit your home!

Timber fences have traditionally been the poor cousins of more expensive brick styles. This has changed quite radically in the past few years as smart designers have begun to exploit the real advantages of timber.

Timber is certainly an easier material to handle than brick. For the do-it-yourself homeowner who wants to do the work, this is a definite advantage. Timber is mostly cheaper in the long run, and the tools required can be found in most basic tool boxes.

The problems of durability have largely been overcome, so long as the correct timbers and treatment methods are used. Modern paints and stains are long lasting and durable, and the methods of application are more simple. These are just a few of the reasons why timber fences are so popular with home-builders and renovators alike.

Below: An ordinary timber fence dressed up with latticework.

Above: Contrasting colours add interest to this timber paling fence and carport.

Above: Timber lattice fence.

Fence Styles

Picket Fences

A timber picket fence may be ideal for your home if it is a more traditional style of house. Many timber yards make specially designed timber components for picket fences. This ornamental fence goes very well with Federation-style architecture and is easy to build. The structure is made up of posts, usually set in the ground at about 1800 mm centres, with two or three horizontal rails, housed or mortised into the posts. The pickets, which come in lots of different styles, are screwed or nailed onto the rails.

If you're using hardwood pickets you should drill holes for the nails to reduce the possibility of splitting. It's best to use only one nail per rail. Hardwood has a tendency to shrink as it dries out. Two nails, side by side can cause splitting. Always use treated nails, either galvanised steel or copper, for exterior work. Unprotected steel nails will rust and leave ugly black stains in the timber, which are impossible to move. Picket fences are usually available in CCA (Copper Chrome Arsenic) treated pine.

Many local timber yards will be able to help you and it's a good idea to shop around for the right style and price. Some areas, have developed a particular style of fence which is unique to that area. If you notice a common type of fencing and detailing in your neighbourhood keep that 'look' in mind for your own fence.

Post and Rail

The most simple design for a fence is the post and rail style. In this style, square section timber posts are fixed into the ground and are either concreted in, or secured by ramming the earth in and around the hole. Horizontal members, called rails, can be fixed to the posts in a variety of different ways. The easiest way is to nail the rails directly to the posts. This, although it is very simple, is not the best structural solution. A better quality fence will be made by housing or mortising the rails into the posts.

The post and rail style is the basis for most other timber fence styles, as any ornamentation is usually fixed to the horizontal rails.

Timber Lattice, Trellis and Fretwork

Timber lattice, trellis and fretwork were very popular in Federation times. This ornate and interesting timber work has been enjoying a resurgence. Lattice comes in many varieties, but the principles behind its use are standard. Generally, lattice and the other styles are used as infills within a frame.

Lattice works equally well as infill for gates, screens, pergolas, as well as fences. The usual method is to build a frame, out of 100 mm × 50 mm timber, and to fit the lattice inside, securing it with small timber beads fixed to the inside of the frame. As with trellises, lattice may be used with the battens vertical and horizontal, or it may be used on the diagonal.

Lattice is usually made from treated radiata pine or cedar. For locations where future repainting may be a problem, it's better to leave the timber in its raw state, rather than staining. The light greenish grey of the CCA (Copper Chrome Arsenic) treatment is quite pleasant, or if you prefer, use a timber stain finish. CCA treated timber will cause corrosion of unprotected steel fixings so you should use galvanised bolts and screws.

Lattice is a very versatile material, and is used for all kinds of screens, shade structures, as well as fences and gates. Lattice is also ideal for attaching to walls for climbing plants.

Tea-Tree Fences

Tea-tree fencing is another old favourite which is enjoying a new popularity because of its informal look and the way it suits Australian native planting.

Tea-tree is usually supplied and installed by specialist fencing companies. It is reasonably economical and long lasting. Good results can be achieved when tea-tree is used with timber and brick. Very attractive fences can be made using a brick dwarf wall and brick piers, with tea-tree infills. As the wood tends to bleed sap for a short time after installation, it's a good idea to delay the painting of any brickwork or render, for a few months.

Above: Detail of a tea-tree fence.

Metal Fences

Metal fences are often used for swimming pool security fencing. Galvanised or painted steel and aluminium are normally used. The traditional cast iron, given new life by reproduction in cast aluminium, is also popular.

Tubemakers have introduced a range of prefabricated steel tube fence products for the handyman and expert alike. A variety of styles and colours are available and you should contact the Tubemakers sales office for a catalogue and details.

With new metal painting techniques such as 'Colorbond' roll-formed steel products are being installed where traditional timber fencing once monopolised the market.

Above: Modern style metal pool fence.

Above: Painted metal front fence and entryway.

Above: Metal pool fence.

Pool Fences

Security fencing for home swimming pools poses a special set of problems. Due to the high number of tragedies involving small children drowning in pools, fencing is required by most local councils. The result, in many cases is an ugly, high fence which divides your garden. Often the fence reduces the amount of leisure space around the pool and can create pockets of useless space.

The Standards Association of Australia has issued a set of rules for the design and installation of pool fences. Check these standards, as well as local variations, at your local council offices.

Safety requirements vary from council to council. The basic principles are that all gates must have child-proof locks and that the fence must have no hand or foot holds for children to climb up and gain access to the pool. The problem of pool fencing can often be resolved at the pool construction stage by choosing the right location. Boundary fences are acceptable under the safety standard, as well as walls of the house, as long as there are no doors. If you're lucky, that will leave you with only one or two sides to fence.

As the area around the pool is corrosive, due to the chlorine in the water, any metal fences have to be well protected against rust. Galvanised steel, aluminium and colorbond steel are the metals commonly used for pool fences.

Major manufacturers of steel and aluminium fencing have developed special designs for pool fences which comply with the safety requirements. Check with your local building information centre for product displays and information brochures.

Above: Brush fence bordering pool area.

Below: metal pool fence surrounding a beautifully landscaped, multi-level pool area.

Above: Colourful oriental style fence and gate design.

Left: Entry gate and portico designed to blend into surrounding greenery.

Above: Victorian style entry gate and portico.

Above: Contrasting modern style painted timber fence.

Above: Metal entry gate to pool area.

Gates

When someone arrives at your home they should feel welcome. The front gate must be secure, but should also convey the feeling of being 'open' to visitors. The design and details of your entrance gate will be an important part of your home's overall style. You may prefer to have a metal gate that people can look through to the garden beyond. Other homes have sturdy timber gates that give a feeling of strength and security. Timber gates are easy to build provided a few simple tricks of the trade are observed. Gates are usually treated roughly, so your design needs to be sturdy and substantial.

The simplest timber gate has a rectangular frame around the perimeter, to which the facing is attached. The key to successful gate construction is in the design of the diagonal bracing. All timber gates must be braced to prevent them from 'sagging'. The diagonal brace should be mounted as close to the lower hinge as possible, and extend to the catch, lockset or latch of the gate. The brace should be 'in compression' not in tension.

As all external gates are vulnerable to the weather, only use galvanised or brass fittings, and use an exterior grade lock. When metal fittings rust it causes black stains in the timber, and even decay. If you don't want to make your own gates, look at the standard designs stocked by the larger builders suppliers. If you choose a standard design make sure the opening is suitable — it can be very expensive to have special sizes made up for non-standard openings.

Metal Gates

The most common materials used for metal gates are galvanised steel and aluminium. Traditional styles of cast iron fencing are being reproduced in aluminium designs which suit older style homes.

Many small companies specialise in metal gates and security grilles. Look in the *Yellow Pages* for a local metalworker. As most metal gates are made to order, you can design exactly what you want. Most suppliers have standard models or will assist with any special design.

Above: Letterboxes with house numbers are an important part of any fence design.

Letterboxes and House Numbers

It's easy to forget about placement of your house number and letterbox when installing a new front fence. If your house number is mounted on the front of the house, make sure it's not obscured by the fence. In case of an emergency, it's essential to have your house number clearly displayed and well illuminated at night. Many local councils will make this a condition of approving your plans.

It's important that your letterbox is well placed in relation to the fence, as well as being in the most practical location for its purpose. If your driveway runs through the front fence, you will have to return to the footpath to collect mail. If you park the car on the street, the letterbox should be at the front gate. Try to place the letterbox on the same side of the entry door as the door handle as this makes access to the letterbox from the inside of the house easier. (This sounds simple, but is a point often overlooked at the design stage.) Be sure to install a good-sized letterbox. There's only one thing worse than junk mail in your letterbox, and that's when it blows all over your garden and front lawn!

Making a small, timber letterbox is a pleasant afternoon's work, and doing it yourself is a great way to save money. Alternatively there are many suppliers of prefabricated letterboxes of all shapes and sizes designed to suit both timber and brick fences.

Guidelines for Building a Timber Fence

The following is a description of a simple design for a low front fence: 100 mm dressed hardwood posts, mortised for 75 mm × 50 mm dressed hardwood rails, with traditional colonial palings, spaced at about 25 mm.

Make sure that you don't build your new fence over the boundary of your property. Check the original survey. This should show the location of the corners. The surveyor may have put down timber pegs to indicate the exact position of the corners. If you're replacing an existing fence, check to see that the original one was correctly placed and mark out the new one accordingly.

If your home is the first one in a new block, you may have to employ a surveyor to mark out the corners of your site. In most cases, however, you'll be able to locate the correct position without problems. When you have located each corner, mark its position with a pair of timber stakes, and run a tight string line between them. Once you've done this you'll be able to set the positions of the post holes and start digging.

Depth of post holes for fences will depend on a number of local factors. The type of ground encountered will be the most important one. In soft sand the holes should be deeper than in stiff clay or shaly ground. As a guide: a hole for a post that will be less than 1200 mm high should be around 600 mm deep, while a full height fence 2400 mm high, would need a hole about 900 mm deep. Allow a little more for sandy soil.

Common practice varies from location to location. Some fencers use concrete, while others prefer to ram home the soil, and rely on the strength of the soil to resist wind loadings.

Pre-cut and mortised timber fence posts in a variety of styles and heights, are available from most timber yards. This can significantly reduce the labour in building a fence. The rails of the fence can be housed into the posts, or inserted through mortises. A mortise is a rectangular hole cut through a piece of timber. The cutting of a mortise is time-consuming if you have to do a number of posts. The work is doubly difficult if you're working in hardwood. If the commercially available posts can be used in your design, the small extra cost of ready-made posts is worth it.

The other common method of attaching posts together, is to cut a housing joint in the post. This will normally be at least half the thickness of the rail, or as much as the full depth.

An essential tool for the job of fence building, as with all carpentry work, is a carpenter's level — the longer the better. Combined with a string line, a carpenter's level will help in the positioning of the fence posts and rails. Careful work is the very basis of successful building. When you are the builder, time is relatively cheap and the extra time taken to get it right, costs next to nothing.

Building materials can be very costly so it's worthwhile taking special care of good materials.

Setting Out

Once you have established the boundaries of your property and pegged out the corners, you can accurately set out the position of the fence.

Drive a pair of 'sight boards' over the boundary line and position a nail on, or slightly inside the line. It's sensible to position your fence a little inside the boundary line so that any small inaccuracies don't result in your fence falling outside the line.

Project: BUILDING A SIMPLE TIMBER FENCE

The basic principles for building a timber fence are the same, regardless of the design and decoration. All timber fences have posts set into the ground and horizontal rails onto which the fencing material is fixed.

Most timber yards have pre-cut posts and other fencing materials. Posts can be purchased with the holes for the rails pre-cut. The holes are called mortises.

The success of any timber fence is dependent on the care taken when the posts are erected. You can rely on rammed earth to support the fencing, or use concrete around the post bases for additional support.

Use a string line to set out the boundary and carefully excavate for the posts. Use temporary bracing to keep the posts vertical during erection.

Fill the base of the holes with some crushed rock, like aggregate, to assist drainage. Treat the bases of your posts with a timber preservative like creosote.

The joints in the horizontal rails should be cut on the mitre and be located at a post.

The kind of palings or infills you use are a matter of personal taste and fence style.

TOOLS REQUIRED
spade, shovel, pick
stringline and level
power saw, hand saws, drill hammer

Note: An old sagging fence can be resurrected with a length of 75 mm × 75 mm × 10 mm galvanised steel angle driven into the ground beside the post and bolted into sound timber.

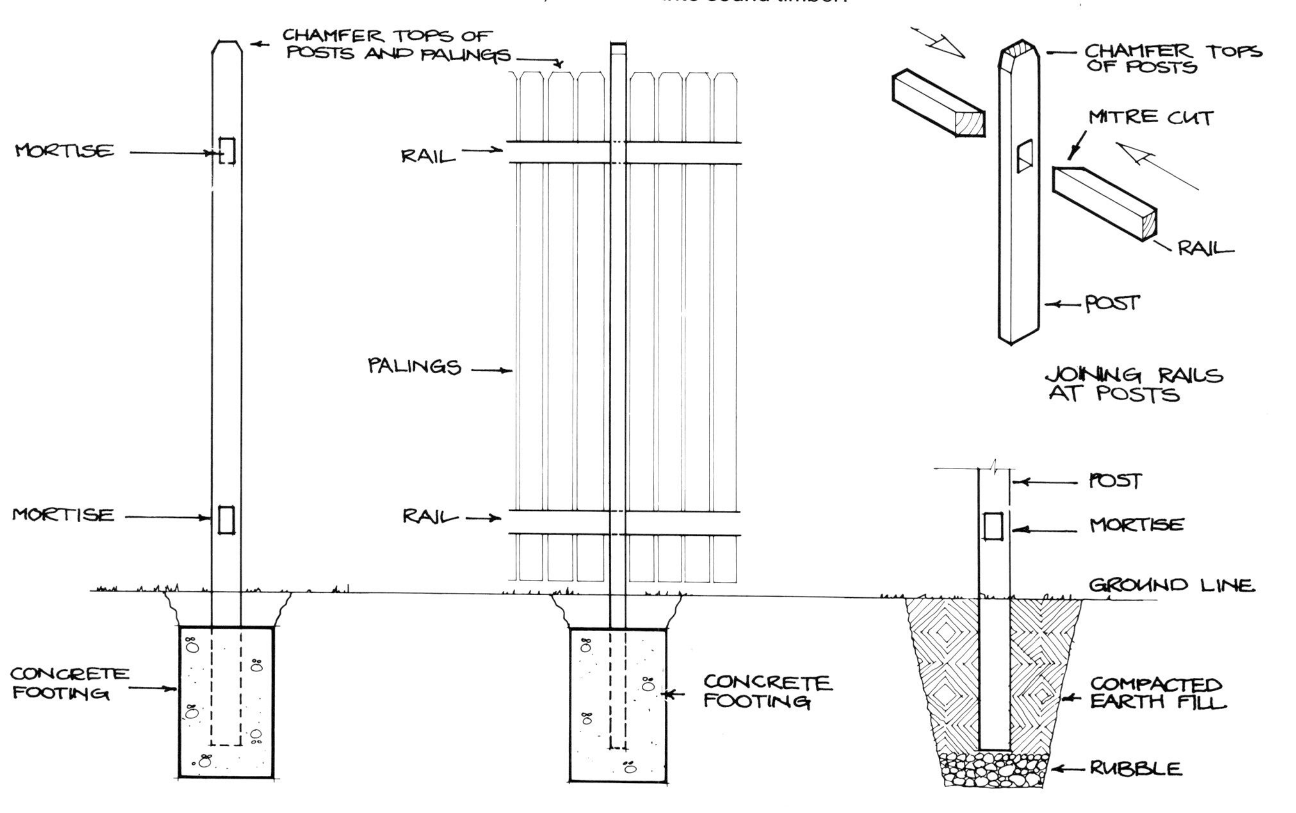

7

SHADE STRUCTURES: PERGOLAS AND GAZEBOS

Pergola Design • Pergola Construction • Footings • Structure, Joints and Connections • Bracing • Covering • Project: Building a Pergola • Gazebos • Guidelines for Building a Gazebo • Project: Building a Gazebo • Types of Timber • Finishes

Pergolas have been described as 'horizontal trellises or frameworks supported on posts that carry climbing plants and may form a covered walk'. This definition can be extended to include most shade structures connected partially or fully to a house.

Modern pergolas serve a dual purpose, first as supports for climbing plants, and secondly, as highly effective passive solar screens. With the help of deciduous plants, a pergola can screen out the worst of the sun from spring through to autumn, and in winter, gives maximum sun penetration.

In this book, the term 'pergola' has been used to cover a broad spectrum of shade structures, including arbours, pergolas and, to a lesser extent, gazebos. The construction techniques and design principles for pergolas also apply to decks and terracing (see Chapter 3). While there are subtle differences between the various types of shade structures, the basic carpentry involved is the same.

A gazebo, like a pergola, is designed to provide a good degree of shelter and shade. With its characteristic hexagonal or octagonal shape and segmented roof, a gazebo also provides a delightful additional place for al fresco family meals and entertaining in summer.

From top: Two views of treated log pergola with wire for training plants.

Below: Pergola designed to provide poolside shade.

Pergola Design

The design and construction of a pergola is probably one of the most simple do-it-yourself projects there is. The joinery work is straightforward and you should be able to build a pergola as a weekend project. If you don't feel confident about your carpentry skills, get a quote from one of the many companies who specialise in pergola design, construction and installation.

This section of the book looks at the basic principles of pergola construction. Whether you decide to build one yourself or hire a professional, these design hints and carpentry basics will act as guidelines to achieving a really professional result.

Council Approval

A pergola is, technically, a 'structure' and you will need to apply to your local council for permission to build. Some councils may consider this formality a nuisance, however it's always a good idea just in case there are any objections. Contact your local building inspector and discuss the details of your design. The council is responsible for safeguarding not only your interests but also your neighbours'. Cooperation with both council and your neighbours will prevent a lot of possible strife! When you design any outdoor structure you must consider how it will affect your neighbours.

A well-designed pergola should block out all the sun when heat and glare are a problem, and let maximum winter sun in to warm the interior. A number of methods are used to achieve this result. The most successful pergolas use a combination of devices for the best effect. Sunlight can be controlled by the spacing and size of the rafters, with shadecloths, translucent sheeting or closely spaced timber battens, or even with deciduous vines and other climbing plants.

The design you use for your home will depend on a number of factors.

Orientation: north, east or west facing, or a corner
Position: views, where the problem sun comes from
Location: the pergola may be free-standing or attached to the house
Aesthetics: the design has to suit your home

Basic Principles

We all know that the sun rises in the east and sets in the west. Unfortunately it isn't that simple.

The sun is directly overhead at midday on 22 December for someone standing on the Tropic of Capricorn. In midsummer the sun will rise in Sydney almost due south-east, which is 45 degrees further south than east. On the same day, the sun will set nearly due south-west. For Sydneysiders the sun will never rise to an angle of 78.5 degrees above the horizon.

In winter, the sun will rise in Sydney about east-north-east on a bearing of about 65 degrees, and set about west-north-west or 295 degrees. Its maximum altitude at noon will be about 33 degrees. A working knowledge of these astronomical facts is essential when designing a pergola.

Winter Sun, Summer Shade

In summer your pergola will need to provide generous amounts of shade for outdoor seating areas and the interior of the house. In autumn the sun starts to drop towards the north, and the amount of sun penetration into your windows increases.

As winter approaches, maximum penetration should occur, providing pleasant conditions for outdoor eating and entertaining. The additional sun penetration into your home will help warm it and save on heating bills.

Plants

The perfect pergola has yet to be built due to the very different demands of winter sun and summer shade. Many designers use deciduous plants as an additional control element. Deciduous vines like wisteria and ornamental grape vines have dense foliage during the hottest part of the summer. The dense leaves give a wonderfully filtered light and plenty of shade from the burning sun.

As the autumn days grow shorter, and the sun loses its heat, the deciduous plants begin to lose their leaves. The light and heat progressively increase to light and warm. In winter when you need all the warmth and light you can get, the vines are bare.

A well-designed and soundly constructed timber pergola, covered with deciduous climbing plants or vines, is the ideal shading device.

Left: These climbing plants provide delicately filtered light in spring and summer when it is most needed.

Below: Bougainvillea adds a splash of colour to this simple pergola.

Above: Pergola enclosing a small, private courtyard.

Left: Light, airy pergola extending over a deck.

Pergola Construction

A pergola is made up of a timber frame supported off posts, which have to be connected to a footing. Sometimes the footing will be an existing concrete path or slab, but usually you will need to make a concrete path to take the load.

Pergolas can be free-standing (that is, unattached), but are usually attached to a wall. They also need some form of covering — with shadecloth or the shading effect of rafters or battens attached to the frame.

There are five basic elements in pergola construction: footings, structure, connections, bracing and covering.

Footings

There are different methods of fixing timber posts into the ground. The more traditional method involves burying the end of the post in the ground. Some builders use concrete to surround the post to further secure it, while others place a bearing pad at the bottom of the hole, to take the weight, and ram the earth to compact it.

Any method whereby the end of the post is buried in the ground, can create future problems, such as exposing the timber to ground water and insect pests. You may not notice any damage until it's too late. If you decide to use this method, make sure that your drainage is adequate and that the timber has been treated to resist rotting and insect attack.

Concrete pad footings are the most popular method of securing timber posts into the ground. The wide variety of commercially available metal shoe fasteners makes the detailing of the base of a pergola post simple. The many different styles of post anchors discussed here are all available from your local building supplier.

For a typical pergola, a concrete pad footing of some 350 mm diameter and about 450–600 mm deep, will be adequate. Some formwork will need to be constructed to contain the wet concrete until it has set. Waterproof plywood, sometimes called 'formply', is what the builders use. For this type of work however, any timber will do.

One very effective trick is to use an old 20 litre paint drum with the ends cut out, as the formwork. The steel tube is placed in the excavated hole and the hole backfilled against the form. It is essential that the new concrete is placed on well-compacted soil to avoid any settlement as the footing is loaded.

If you need holding-down bolts for your post anchors they can be cast in when the concrete is poured.

Structure, Joints and Connections

The choice of timber for your pergola is largely up to you. Oregon or Australian hardwoods are both suitable. Oregon, while being easier to work, is more likely to develop cracks as it dries out which is why the rough sawn grades of timber are more suitable than dressed timber.

Square timber posts for pergolas may be either from 100 mm × 100 mm timber, or, for smaller pergolas, 75 mm × 75 mm. It is particularly important that the timber is straight-grained and well-seasoned, as any warping and twisting is especially noticeable on a pergola.

The foot of each post is housed in the post connectors and the bolt size should be selected to minimise the length of thread exposed as this helps reduce rusting. The technique of counterboring the hole to a depth sufficient to conceal the nut and washer is a good design practice.

The construction joints used for a timber pergola are basic and simple to execute. The most common type is called a 'halving joint'. This joint, as the name suggests, is made by removing a section of timber from one member to allow another to fit into the recess. This joint is often used to support beams over the posts of a pergola.

The more complicated 'mortise and tenon' joint may occasionally be used for the same purpose. The cross member is cut to form a 'tenon' or tongue which fits tightly into a 'mortise' or pocket, cut into the other piece. The most common use for this more elegant style of joint is in the legs and rails of tables.

Modern pergolas use a number of composite joints fabricated from timber and galvanised steel plates and bars. The simplest of these is the joint between the posts of the pergola and the timber beams. Both the post and the beams are recessed to allow the surface of the steel jointing plate to sit flush. (A plate is a horizontal load-bearing member. In pergola design, a plate is required at the house wall and another supported by the line of posts.)

The specially fabricated steel plates can be made by the average builder. They need to be hot dip galvanised to prevent corrosion and rusting. (Galvanising companies are listed in the *Yellow Pages*.) Galvanised steel fabricated plates have almost totally taken over from other methods of attaching timber posts to their footings. It's unusual to see timber posts set in the ground or in mass concrete now because of the convenience of these prefabricated fittings. All hardware and builders suppliers carry a range of ready-made fixings for almost all designs.

For a less conventional design you may need to use specially made flat bar steel fittings. Check the *Yellow Pages* for metalworkers, although some of the larger hardware and builders' suppliers will cut simple steel items for you.

Above: This timber pergola is attached to the house under the eaves line.

The simplest steel bracket consists of a length of galvanised steel flat bar drilled for connection bolts. Four holes (12 mm diameter) allow the plate to be connected to brick walls, cast into footings of mass concrete or used as a connector for two timber members. The plates may be cut from arch bar material measuring 76 mm wide × 10 mm thick, and can be cut to the specified length. Standard lengths, as with most building materials, are in multiples of 300 mm (1 ft). Special lengths can be cut to order.

If you cut the material yourself, remember to bury the end which you cut into the concrete footing, or use a cold galvanising paint to protect the steel from rusting.

Bracing

Your pergola will need some form of timber bracing to stabilise it unless it is built inside two walls at right angles to each other. This may take the form of knee braces between the posts and rails or diagonal braces in the plane of the rafters.

A diagonal brace can be inserted at an angle 45 degrees, fixed between the double beams and attached to the posts. The brace needs to be carefully measured so as to fit neatly between the beams, secured with galvanised nails or galvanised steel bolts, and nailed at the post. When setting out for this brace, the double beams must be housed into the posts sufficiently, so that the space between them will be equal to the thickness of the angled brace.

If you don't want to use braces, the posts can be cast into substantial concrete footings, relying on the cantilever effect to stabilise them. However, this method is best suited to free-standing pergolas and only if treated timber is used.

Covering

The location and aspect of your pergola will determine the kind of covering you choose. For example, a north-facing pergola should filter the higher midday sun. If your pergola faces west it should be designed to control the rays of the lower, afternoon sun. Effective sun control can be achieved by using a combination of suitably spaced rafters and battens, and sometimes with shadecloth covering. The rafters and battens can be spaced close enough to block out the sun at the required time of day.

The horizontal spacing for rafters depends on their height. If you use 100 mm × 38 mm size rafters they should be spaced about 400 mm apart. For rafters 150 mm deep the spacing should be about 600 mm. It's important to take into account the times of day when the pergola will most likely be used.

A decorative covering can be created by using small diagonal battens, from timber 50 mm × 25 mm, laid over rafters spaced as far apart as 900 mm. This method provides a highly effective sunscreen which is also visually interesting.

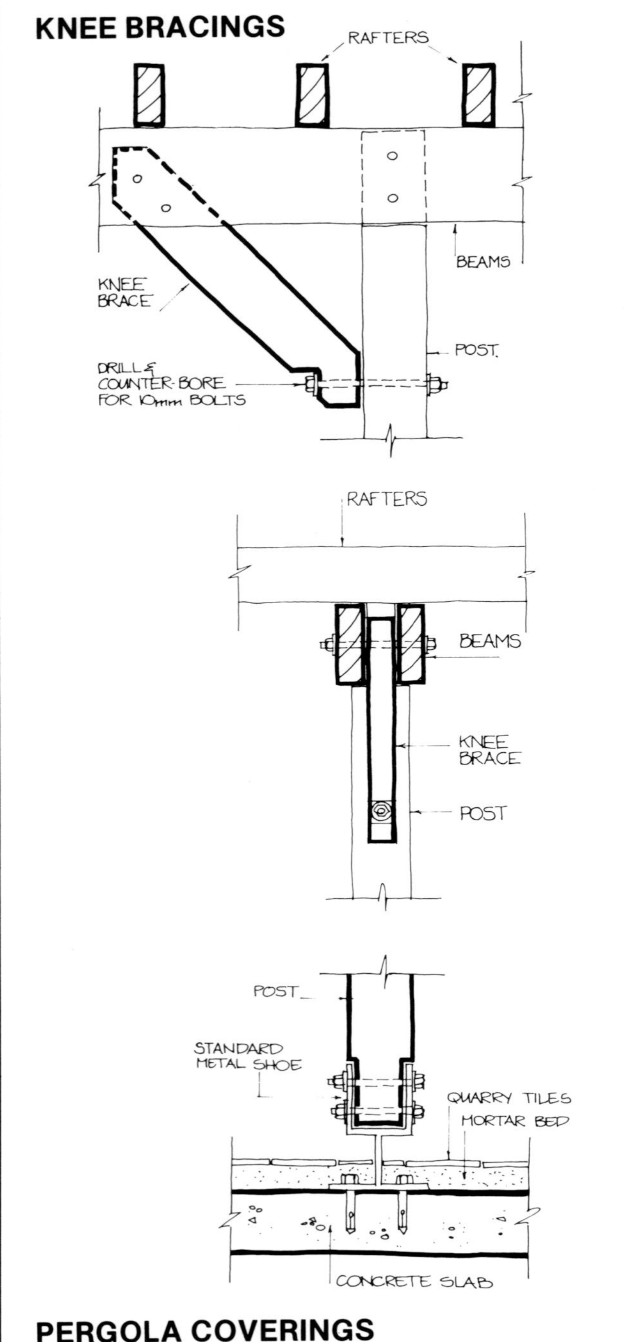

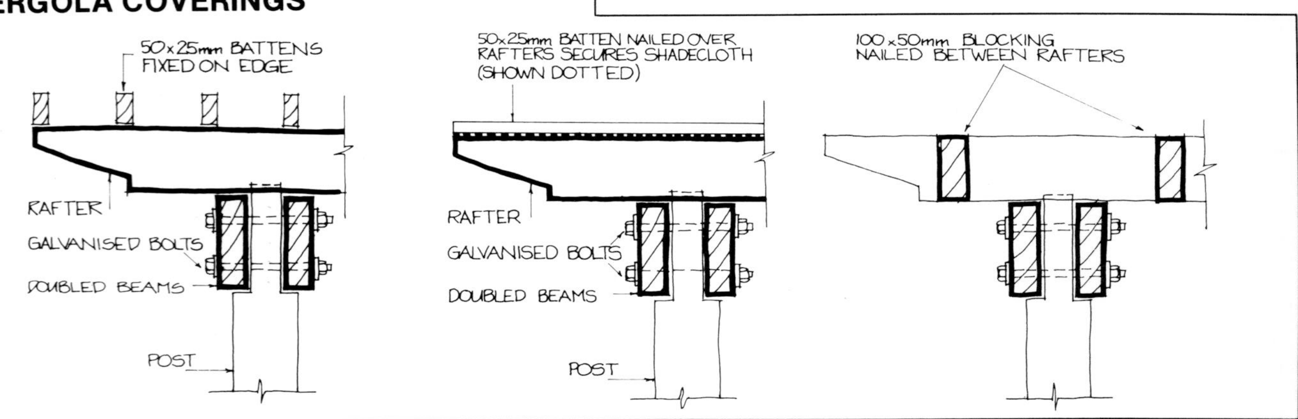

Shadecloths are woven plastic fabrics which come in a variety of densities, depending on the percentage of light transmitted. The most useful shadecloths are in the range of 80 to 90 per cent. These fabrics can be secured to the tops of the rafters with staples or timber battens. In any case, all nails, screws and staples must be heavily protected against rusting, preferably by galvanising.

To reduce the glare and heat of the afternoon summer sun the western end of your pergola can be closed off with a trellis, lattice work or by planting appropriate shrubs.

Project: BUILDING A PERGOLA

A pergola consists of three basic elements: footings, structure and coverings.

If you have a concrete slab you will be able to bolt standard post 'shoes' to the concrete with masonry anchors. Carefully set out the positions of the posts and using a tungsten carbide masonry drill, drill holes into the concrete for the masonry anchors.

If your pergola is part of a larger project, and you have no slab for a footing, it's no problem to make the footings yourself. Excavate for the footings, making sure that the ground is firmly compacted and the surrounding earth is disturbed as little as possible. Use plywood, or a 20 litre paint drum with the ends removed, as formwork to hold the wet concrete until it sets.

Set the top of your footing to allow for the thickness of the paving and the design of the connection you require.

Attaching your pergola to the house is simply a matter of fixing a wall plate to the timber studs in the case of a timber home, or to the exterior brickwork. Use a carpenter's level to ensure that your work is perfectly horizontal.

Rafters span from the wall plate to a bearer fixed to the top of the posts. For ease of construction, use a lapped joint for the bearers. This joint provides temporary support until the bolts are fitted.

Rafters may be horizontal or pitched to a slope, depending on your design. When rafters cross a plate at an angle, allow to 'birdsmouth' them ('V' shape cut out of timber so that the rafter sits snugly across the plate) for a neater joint.

Coverings for a pergola may be in the form of timber battens, timber blocking, lattice and trellis work or shadecloths.

Stop chamfering all exposed edges of posts, bearers and rafters makes an attractive finish. Chamfering timber also makes it less likely to split and splinter.

It is best to stain or undercoat the timber before you erect your pergola. Painting on the ground is infinitely easier than doing it from a ladder.

Counterboring the bolt holes so that the bolt heads lie below the surface of the timber is also recommended. This is especially important for the post connections where exposed boltheads can easily cause an injury. Use washers and galvanised bolts (ensuring that the galvanising is not damaged by overtightening).

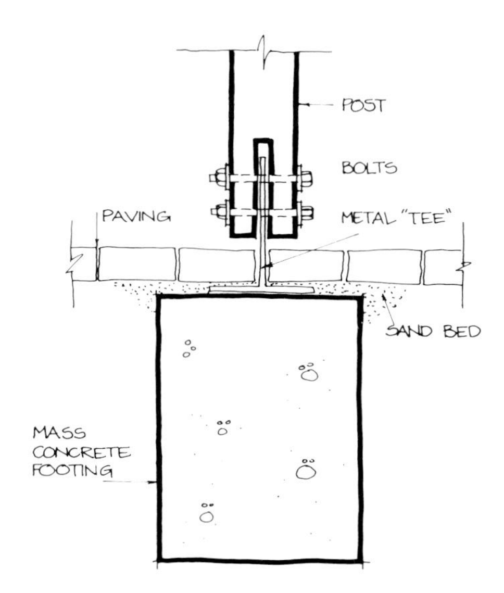

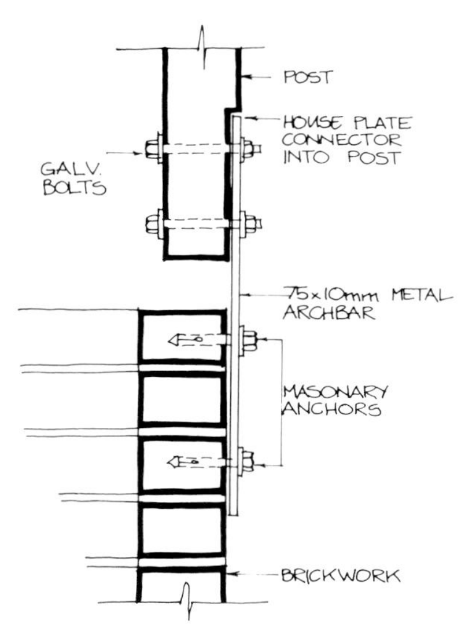

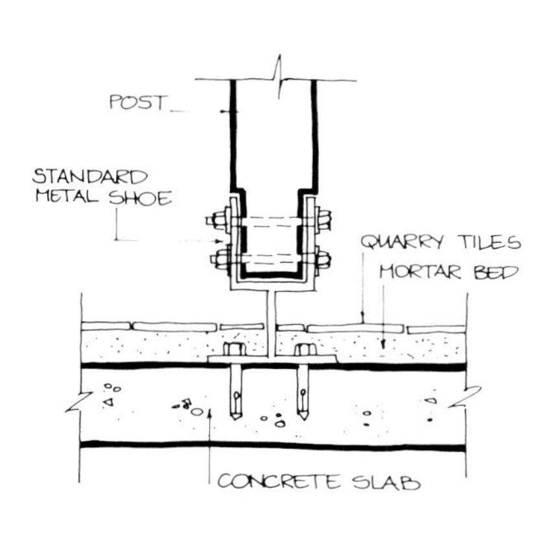

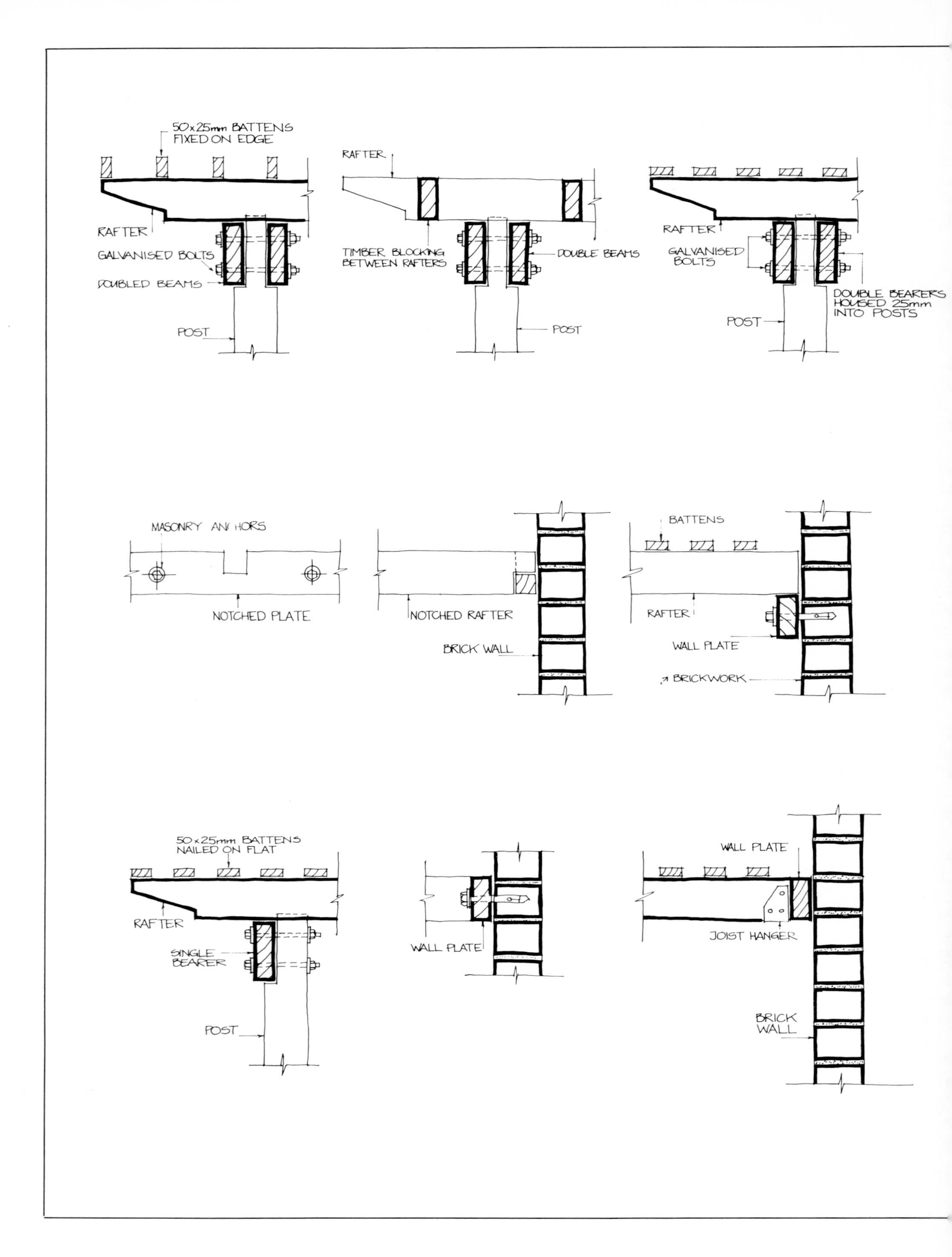

50x25mm BATTENS FIXED ON EDGE
RAFTER
GALVANISED BOLTS
DOUBLED BEAMS
POST
RAFTER
TIMBER BLOCKING BETWEEN RAFTERS
DOUBLE BEAMS
POST
RAFTER
GALVANISED BOLTS
DOUBLE BEARERS HOUSED 25mm INTO POSTS
POST
MASONRY ANCHORS
NOTCHED PLATE
NOTCHED RAFTER
BRICK WALL
BATTENS
RAFTER
WALL PLATE
BRICKWORK
50x25mm BATTENS NAILED ON FLAT
RAFTER
SINGLE BEARER
POST
WALL PLATE
WALL PLATE
JOIST HANGER
BRICK WALL

Gazebos

A gazebo is an outdoor structure designed to provide a degree of shelter and shade, and is normally hexagonal or octagonal in plan, with a segmented roof.

A gazebo is a delightful addition to your outdoor entertaining and family life. It provides an additional place for family meals in an informal al fresco atmosphere, as well as providing shade and shelter for just lounging around.

Gazebos were very popular in Victorian and Federation times, and a number of specialist manufacturers are leading a resurgence in fashion. Because the joinery work required for a gazebo is quite exacting, some companies have begun to market a range of prefabricated designs. They will also supply and install a variety of designs to suit most situations. Buying a kit for your gazebo is a good compromise, as the detailed work, particularly in the roof, can be difficult for an inexperienced carpenter.

Gazebos can be made from timber or from cast iron or aluminium. The modern reproductions of old cast iron designs combine traditional elegance with the advantages of modern materials. New timber treatment processes also provide superior resistance to the elements and insect attack. The basic construction principles are a little different to conventional structures. The angles of the traditional designs make the joinery more complicated and even an experienced carpenter will have difficulty making the joints at the vertex of a gazebo. Structurally a gazebo poses a similar construction problem to a free-standing pergola or arbour.

A gazebo can be square, hexagonal (six-sided), or octagonal (eight-sided), in plan. It's best to make a gazebo with an even number of sides to simplify the timber work. As long as each pair of rafters matches up correctly, the structure will be stable during erection.

The critical component of a gazebo is the hub or 'finial' . It's this piece of timber that supports all the rafters so that the load of the roof is properly distributed. For a six-sided gazebo, the hub at the top of the roof will be hexagonal also. For a gazebo with eight sides, the hub member will be eight-sided. You'll need to make sure that the hub is big enough to safely connect all the rafters together. During assembly, you may need to prop up the hub until the structure is complete.

Timber gazebos can be roofed with any suitable materials, for example, corrugated iron, tiles and slate are commonly used. Complicated roof forms, made from multi-sided shapes, can sometimes make the roofing job time-consuming. Careful and meticulous work is needed for good results.

Above: Victorian style gazebo built on the water's edge.

Guidelines for Building a Gazebo

A gazebo can be as simple or as elaborate as you want. You can build it on a timber base, or on a concrete slab. However, as the gazebo shape can be square, hexagonal or octagonal, it's important that the concrete formwork is set out with care. Standard post anchors made from galvanised steel plates are used to connect the corner posts to the slab. These are then bolted to the finished surface of the concrete or cast in at the time of pouring.

For an all timber gazebo, small footing pads are used to support the bearers. The construction of the timber deck is fairly standard (see Chapter 3), except for the choice of timber. Hardwood bearers and joists are the preferred option due to this wood's superior strength and resistance to rotting. Pressure treated timber is also recommended.

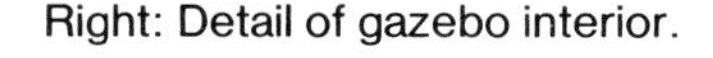

Right: Detail of gazebo interior.

Above: Detail of the finial at the top of the gazebo roof.

The posts of a gazebo are normally made from dressed timber, often hardwood is used. A nice touch is to 'stop chamfer' the arrises or edges of the timber. This is when the chamfer, or planed edge of a piece of timber, is stopped short of the ends or connections. This reduces splitting and splintering, and gives a neat, professional finish.

A plate is attached to the top of the posts to carry the load of the rafters. Erecting the roof framing and the rafters is the only really tricky part of building a gazebo yourself. The secret, as with all carpentry, is to work carefully, always double-checking your measurements.

The top of the roof, where all the rafters come together, is called a 'finial'. In a simple, square gazebo this may be made from square timber, say 100 mm × 100 mm. The finial must be propped until the first pair of rafters have been erected and the whole of the roof framing made stable. For hexagonal and octagonal gazebos, the finial should be round or have the same number of faces as there are sides on the gazebo. Allow a sufficiently large finial for all of the rafters to fully bear on the face of the finial (see Project, page 78).

In more complex designs a star-shaped metal-bracket is often fixed under the finial to connect all of the roof structural members together. A local metal-working shop can make the plate to your requirements from steel plate or flat bar. The normal thickness would be from 6 mm or 10 mm material. The bracket should have the same number of points as the number of sides in the gazebo, and be as wide as the roof members. Each rafter is then fixed in position with 10 mm diameter coach bolts inserted through holes drilled in the connector bracket. As with all external metalwork, the most effective rust proofing is hot dip galvanising.

Above: This rustic gazebo has a shingle roof and creates a charming shelter for plants.

Roofing a complicated shape is easier if you choose the appropriate material. Timber shingles are an attractive and flexible roofing material. They can be fixed onto closely spaced timber battens, or to sheets of marine plywood cut into triangles for each section of the roof. The sheets are cut to shape with an angle grinder or with a grinder blade used in a circular saw. If you can't borrow an angle grinder from friends or from work, they are available for hire from your local builders yard. Painted corrugated iron is also a reasonably simple material to use.

The ridge flashings for each of the segments of the roof are standard sections available from all hardware shops. The only other tools required for the roofing are a pop-riveter to attach the sections together, and a caulking gun for sealing all the joints with a couple of tubes of roofing silicon sealant.

Guttering and downpipes are optional for most gazebo designs. However, standard section galvanised steel guttering is available from all builders suppliers along with a wide variety of specially fabricated junctions, such as 90 degree bends, downpipe connections and elbows, which make roof plumbing simple. The more modern plastic sections and guttering systems also make the job a breeze. If your gazebo is located away from the house and the stormwater system, you may need to direct the water to a drainage pit. As a general rule the size of a drainage pit can be estimated by allowing one cubic metre (that is, a hole 1 metre square by 1 metre deep) for every 25 square metres of roof area. This is dependent on the absorption capacity of your soil and the amount of rainfall expected.

Decoration

Decoration of the basic gazebo frame described above will be determined by your taste and the design of your home and garden.

Most gazebo designs use infill panels between the posts; some use timber, while the more traditional ones use cast iron lace panels set between the handrails, and cast iron fringing around the top. Diagonal timber set between horizontal rails, and the use of timber lattice, are also popular style features.

Because gazebos have a relatively compact and intimate space, the quality of workmanship is more noticeable. Careful attention to small details and the use of special techniques like stop chamfering all exposed timber edges, will guarantee an attractive and professional finish.

Below: This pagoda-style natural timber gazebo is an integral part of the pool area.

Project: BUILDING A GAZEBO

Building a gazebo is a rewarding project, but the detailing requires some experience in carpentry and joinery. The delicate structure needs a careful touch. To get the best results proceed slowly and think out each step.

The design shown in the drawings is for an octagonal (eight-sided) gazebo, built off concrete pad footings and small brick dwarf walls. Concrete footings have been described in Chapter 4, so if you need more information refer to the section on footings.

The joists are doubled to allow the connection of the square posts and to provide a continuous support for the timber decking. Jack joists are connected to the main joists with metal shoe connectors and are supported off small brick piers.

All structural timbers are stop chamfered which is done by planing off the edges of the timber between joints, leaving a section of some 75 mm.

A galvanised steel connector plate in the shape of a star is used to secure the vertex (highest point) of the

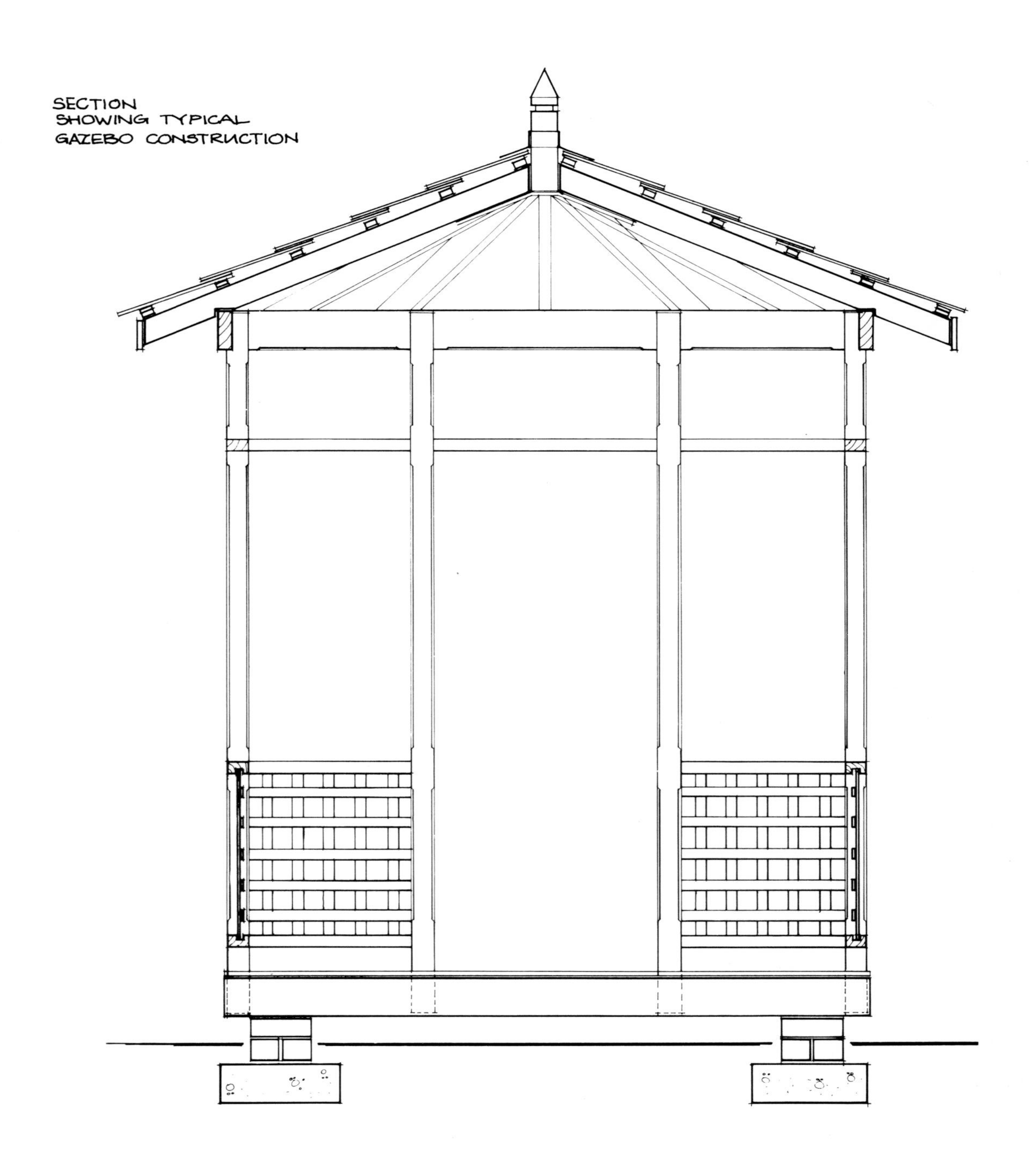

oof. Bolted connections are used where possible. ialvanised fittings should be used throughout for uperior corrosion resistance.

The timber floor is laid on the diagonal to reduce the umber of jack joists and secondary framing. Well-easoned hardwood should be used for the flooring as azebos are exposed to a lot of weathering.

The roof shown is a cedar shingled one. This type of oofing is very popular and has the advantage of equiring few special tools or techniques in its nstallation. Be prepared for some intricate cutting work and some difficult flashings if you want to use a metal or tile roof. In general, the shingle system is well suited to the handyman builder.

Gutters are really optional for an informal outdoor structure; there is a fascia (a flat band or surface) shown on the drawings, and downpipes can be attached to the posts.

Treated timber lattice work fixed inside a rebated frame is used for enclosure. Cast iron lace or more commonly now, reproduction aluminium lacework, is also a possibility.

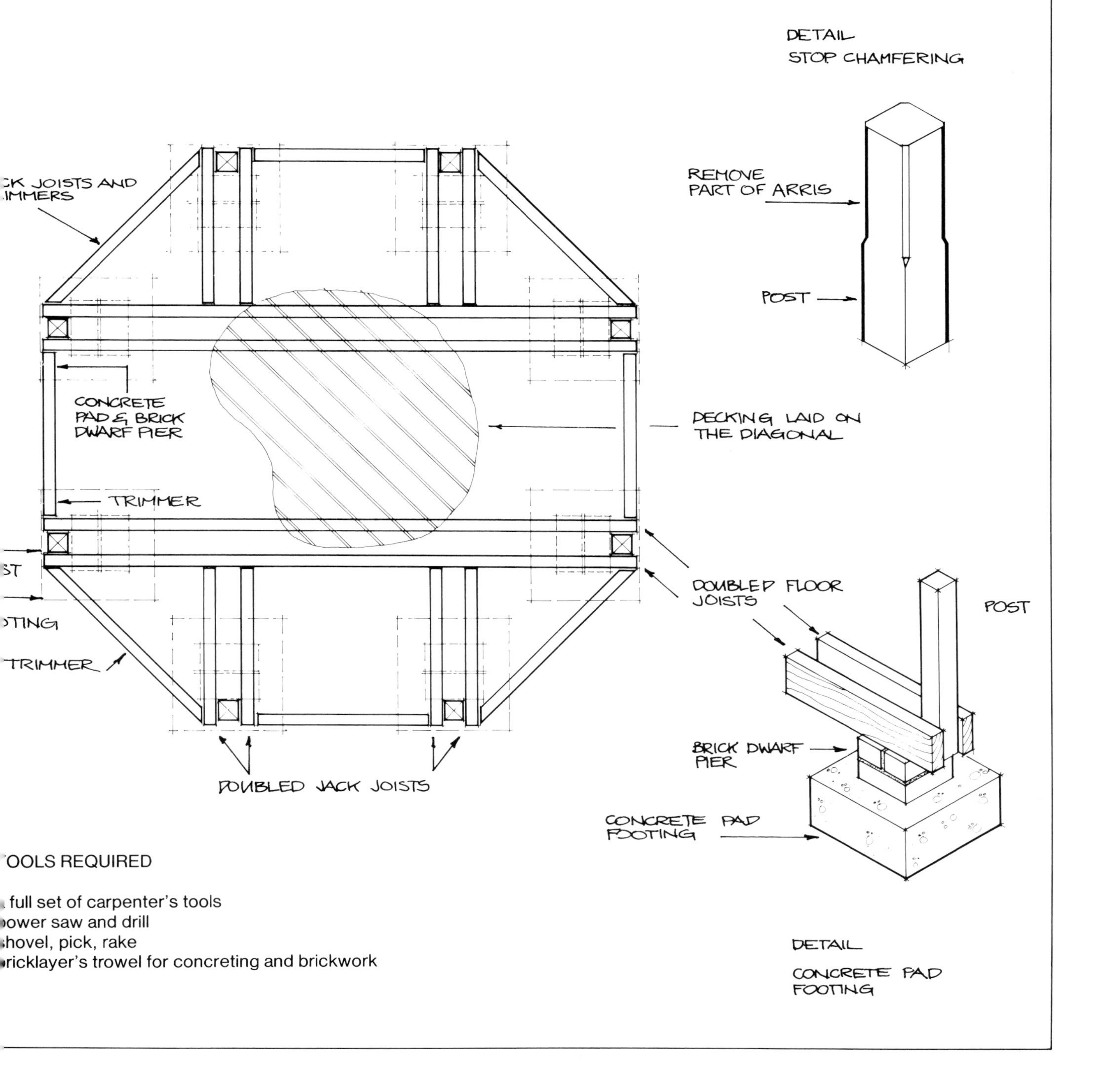

'OOLS REQUIRED

full set of carpenter's tools
ower saw and drill
hovel, pick, rake
ricklayer's trowel for concreting and brickwork

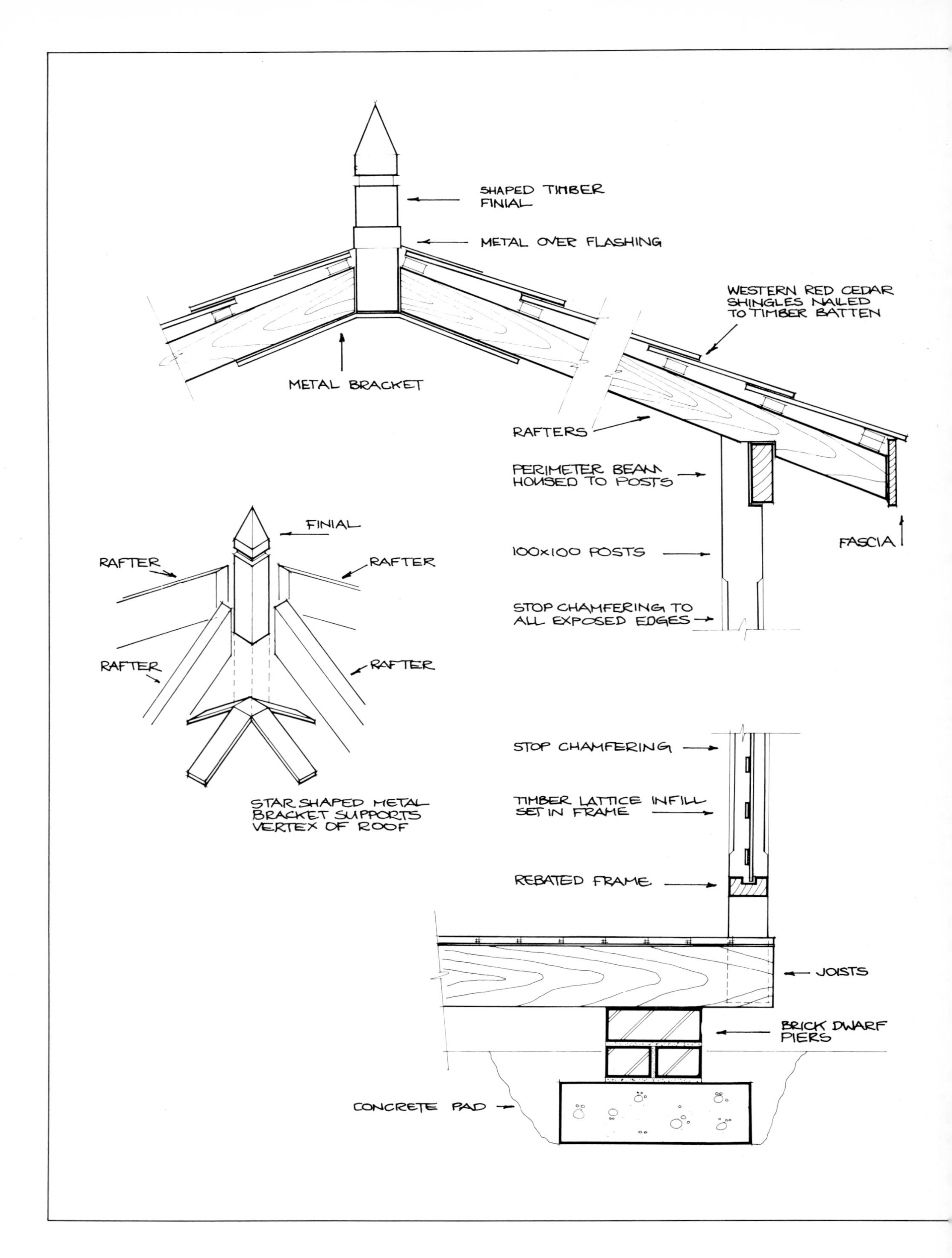
SHAPED TIMBER FINIAL
METAL OVER FLASHING
WESTERN RED CEDAR SHINGLES NAILED TO TIMBER BATTEN
METAL BRACKET
RAFTERS
PERIMETER BEAM HOUSED TO POSTS
FASCIA
FINIAL
RAFTER
RAFTER
100x100 POSTS
STOP CHAMFERING TO ALL EXPOSED EDGES
RAFTER
RAFTER
STOP CHAMFERING
STAR SHAPED METAL BRACKET SUPPORTS VERTEX OF ROOF
TIMBER LATTICE INFILL SET IN FRAME
REBATED FRAME
JOISTS
BRICK DWARF PIERS
CONCRETE PAD

Types of Timber

The timber you select for your pergola or gazebo will lepend on what's available locally. Oregon, or more correctly, Douglas fir, is stocked by all timber yards nd is excellent for outdoor structures that have a ough sawn finish, and will be stained for protection. Some carpenters prefer to use Australian hardwoods or their posts, especially if they are to be dressed or planed smooth.

You can now buy timber which has been specially reated to resist borers, white ants, termites and other nsect pests. Some of the manufacturers guarantee imber which has been treated with their product, for s long as thirty years.

The basic treatment method involves the timber eing pressure-impregnated with certain chemicals. You may have noticed the distinctive greenish oloured logs which are used in playground construcion. These logs have been treated with a copperased solution which gives the timber an exceptional veather and insect resistance.

It's important to note that some of these chemicals re potentially dangerous. Don't use offcuts for arbecue fuel. Any offcuts from preservative-treated imber should be disposed of.

The Timber Development Association has a numer of useful pamphlets and information sheets full of leas and hints, for anyone designing and building in mber. A visit to the Association offices or to the local uilding information centre is an excellent idea if ou're planning any type of building or home nprovement project.

Finishes

Because your garden structures will be exposed to the veather, it's important to consider the type of paint, tain or other protection system you want to use. Make a decision early on, rather than at the end of our project.

There are three different strategies for timber proection:

- Timber stains
- Paints and other opaque finishes
- Timber preservative systems

Each system has its advantages and particular appliations. The type of timber and the basic pergola esign will also influence the style of finish that you pply.

Traditional finishes have been extensively mproved in the last few years, and given correct pplication techniques and good quality materials, hould have a service life of more than five years. It's vise to buy the best quality materials that you can fford. There are cheap paints, and good paints, but here are very few good *and* cheap paints. With paint nd stain finishes the cost of the materials themselves s only a small proportion of the cost of labour to apply, plus materials. The few dollars saved by using low quality, cheap materials will be lost many times over when it becomes necessary to recoat, or worse, rebuild.

Stain Finishes

Transparent stain finishes are designed to colour the timber while allowing the grain to show through. Stains come in a wide variety of colours. Most stain finishes have a preservative action and will help resist rotting and decay. Some have the added advantage of surface protection against insect pests.

Stain finishes are sensitive to the surface condition of the timber. If you intend using this style of finish, it should be applied to new timber as soon as possible. As the new timber weathers it dries out and slowly turns a greyish colour. The stain will be absorbed in different amounts according to the condition of the timber and can give an undesirable mottled effect. Because the surface condition of any timber will deteriorate, the stain finish should be applied as soon as possible. For practical reasons, it's often better to apply the stain before assembling the structure.

Stain finishes are versatile, but best suited to rough sawn timber. Application can be messy as the stains are not thick like paint. It's a good idea to stain as many of the pergola's parts as possible on the ground before you erect it. In this way the extra work of painting overhead and from a ladder is avoided, and the quality of the job will be better.

Stain finishes are easier to recoat than conventional paints, as there's usually no need to remove the damaged or worn out paint. Remember, however, that once vines and other plants have taken hold, it will be difficult to repaint. Some paints and stains can damage delicate plants and extreme care is needed when repainting.

Paints and Other Opaque Finishes

The modern acrylic paints, as well as oil-based paints, are being used increasingly for external timber work. Generally, the opaque finishes are used on dressed timber, rather than rough sawn timber. Good effects may be achieved with acrylic colours on rough sawn timber.

Conventionally coloured paints are well suited to older styles of architecture. Smooth dressed timbers and well-designed carpentry are complemented by plain white, gloss paint. The effort of recoating, and the eventual need to scrape off old paint however, should be considered carefully.

The modern acrylic paints weather very well and can be recoated without too many problems. They come in a wide range of colours and are highly resistant to weathering. Some manufacturers have specially designed ranges of paints for painting external timber work. Check the colour cards and data sheets at your local hardware shop.

8

KIDSTUFF

Do-it-yourself • Safety First • Materials • Playground Design • What to Build • Involving the Kids • Guidelines to Building Play Structures • Project: Building a Cubby-House

It's a sad reflection on modern life that local councils, aware of insurance problems and their liability for injuries, are beginning to remove all but the most simple playground equipment from adventure playgrounds.

It's also unfortunate that we can no longer assume our children will be safe in public playgrounds. In many areas, playgrounds have become threatening environments for children as a result of vandalism, graffiti and urban crime.

For many parents, an attractive solution has been to build playgrounds at home. It's likely that home adventure playgrounds will become a significant growth industry, as more and more parents realise the advantages of providing their children with their own playground equipment. At least when your kids are playing at home under your supervision, they'll be safe and well looked after.

Above: Commercial design playhouse with tan bark ground cover underneath.

Below: This log playhouse features a balcony and access ladder.

Do-it-yourself

If you can provide a secure and exciting environment for your children at home, then you'll have the best of both worlds. The kids will be happy and you'll be able to carry on with your own activities as well as keep an eye on them.

The play structures you build and maintain yourself can be just as good as the ones in public parks. For those homeowners who don't want to design and build their own structures, there are a number of specialist suppliers who provide an installation service as well.

Prefabricated kits for a huge variety of play structures are available from a number of companies, one of the largest of which is Koppers. The specially treated timber is designed for outdoor exposure and will probably still be serviceable for your grandchildren. The kits can be combined in dozens of ways to make your own unique structures. The best way to be sure that your play structure is a success, is to involve the kids in its design and construction. Most children are capable of banging in a few nails, wielding a paint brush and holding onto the end of a piece of timber, to help.

A play structure should not be too specific. Kids need springboards for their imagination. Even very simple structures can provide a vast range of play opportunities. One day the timber platform will be a fort, the next day it will be a space shuttle or a pirate ship. Observe your children. Often they will enjoy a couple of cardboard boxes more than an expensive, more sophisticated toy.

Safety First

Safety is always a difficult issue, because as well as safety, children need challenges to learn about the world and themselves. Try to design the unit to be as safe as possible, yet still interesting and even educational. The first and most obvious consideration is the ground cover under the climbing structure. If someone does fall, it should be onto sand or some other surface, such as pine bark shavings. A climbing frame over concrete is asking for trouble!

Apart from providing protection if there is a fall, a primary design consideration should be accident prevention. For example, the distance between the treads of stairs should suit 'the average kid', that is, the age group most likely to use the structure. The connections must be carefully constructed, with no sharp edges or splinters. Swings should have generous clearances and be given rigorous weight and strain checks for safety. Hand-holds should be closely spaced and designed for little hands. If your play structures will be used by kids of all ages, then the critical dimensions should be determined to a greater degree by the smaller children. They won't be able to climb up a step designed for a twelve year old.

Some design considerations can be quite subtle, but the safety aspects are very important. The best commercial units have solved these problems with elegance and skill and if you intend to build your own play structure, look at the commercial range for ideas.

Materials

Simple play structures such as swings, platforms and slippery-dips can be easily designed and built by anyone with a very basic set of tools. As long as you have a saw that will cut straight, a power drill that can drill a 13 mm diameter hole and a few garden tools, like a shovel to dig the footing holes, then you can achieve the same results as professionals.

There are some things, like the sliding surface of a slippery-dip, that need some careful consideration. Ideally the sliding surface should be made from stainless steel or fibreglass. Most commercial slippery-dips have a steel support under a thin sheet of stainless steel for economy. Stainless steel is quite expensive, but worth it, if you want your play structure to last. Often the best solution is to buy the commercial unit, and to incorporate it into your own design.

For wooden play structures the best timbers to use are dressed hardwood, treated radiata pine, and round logs. Many of the commercial kits use logs. The connection methods for log structures are not difficult. The section on the construction of pergolas covers carpentry for outdoor structures (see Chapter 7).

Playground Design

The location of play equipment needs to be carefully thought about so that you can supervise your children from the house. With this in mind, select a location which is within eyesight of the kitchen or main living areas of your home. It's a bonus if the play structures get afternoon sun, as the kids will spend a lot of their time out-of-doors after school.

Mature trees (if you're lucky enough to have them) can form an interesting part of a play structure. A tree-house is usually one of the most popular play structures. Be careful to allow for the future growth of a tree when you build near, or attach structures to it. A tree-house is an 'organic structure', in that it must coexist with the tree. Protection of the tree should be an integral part of your design. Avoid nailing directly into the tree as much as possible, and never use copper nails.

What to Build

There are many basic designs which have stood the test of time. The most popular of all structures, the classic cubby-house, can take many forms, whether on the ground, elevated on a platform or in a tree.

Most children love to have a place of their own. The cubby-house is a private domain for a child — a

Above: Timber lattice playhouse with slippery-dip and sandpit.

Right: Tree-house with metal slippery-dip and covered sandpit.

Below: Elaborate two storey tree-house built around two trees.

place to keep precious treasures, and to enjoy time away from parents.

Some children are happy with a large cardboard carton, and can play with it until the cardboard falls to pieces. Others have elaborate tree-houses, some of which are passed on from generation to generation.

Swings and slippery-dips are old-time favourites. The modern designs incorporate slides into adventure structures of quite impressive complexity. The slide becomes part of the escape from the structure rather than a simple free-standing item. Commercially available slides and swings can be combined with forts and climbing frames to provide additional interest.

Another old-time favourite is, of course, the sand pit. Even the pile of sand that you use for bricklaying and concreting is an irresistible attraction to kids. Unfortunately the neighbourhood cats are likely to be similarly attracted, and a well-designed sand pit must have a cover. A sand pit can be made to fit the available space or can be part of a border or pathway.

One simple way to make a sand pit is to use the specially designed split log borders which are sold in bundles. Cut away the grass and other vegetable matter and lay two sheets of plastic underlay. Provide drainage in the plastic sheet, around the perimeter only. Dig the base lower around the perimeter to assist drainage. The border logs may be inserted into the ground and hammered in. About 200–300 mm depth of washed beach sand or plasterer's sand is ideal. The cover for the sand pit is important to stop the sand from getting too wet, as well as discouraging the local cats from using your sand pit as a toilet!

Below: An old tyre and a length of rope make a great swing.

An old tyre and a length of chain or strong rope is enough to provide children with hours of fun. The swing can be connected to a pergola or hung from a larger tree. Make sure you attach your swing to the tree without damaging the bark, and allowing sufficient room for the tree to grow in the future.

Pine bark shavings or sand are the two materials normally used to cover the ground area beneath play structures. A climbing structure, where there is any possibility of a fall, must be designed for maximum safety.

Above: Another use for an old car tyre.

Below: Climbing frame fabricated from treated timber logs.

Involving the Kids

The best indication of the success or failure of play structures will be whether your children continue to use them. One way to ensure this is to involve them in the design and construction.

Your children will be much more attuned to what makes a good play structure work than you are. If they have a definite preference then you should respect their judgement and, within reason, give them what they want. Most parents have had the experience of bringing home an expensive toy which receives a 'ho-hum' reaction. If you don't involve your children in their own play structures the same may apply.

The best play structures combine several elements, such as ladders, platforms and slides, as well as swings. Decoration can be relatively minimal as the children will use their own imagination to make the structure what they want it to be. Successful play structures don't need to be large or sophisticated, a space as small as 2 m square (a bit more than 6 ft square) is enough.

The simplest design can incorporate a basic platform, a climbing frame or ladder, a slide to exit and a swing made from an old tyre. With the right materials this can be built in a couple of weekends.

More ambitious designs, with additional features, may resemble one of the commercially available designs intended for public parks. A large play structure takes a while to build. It can be broken down into manageable sections, and may take several years to complete as more features are added.

Above: Small, low play structure designed for younger children.

Above: Treated timber tree-house designed to last.

Guidelines for Building Play Structures

There is nothing particularly difficult about building your own play structures. The carpentry involves only a few basic joints and no special tools are required. The techniques are similar to those outlined in Chapter 7. However, the materials used for building play structures are usually different.

Square section timber, either dressed or rough sawn, is used for pergolas, while the average play structure will be built from round timber logs. These treated timber logs are a comparatively recent innovation. The distinctive green colour is the result of a special treatment given to the timber to protect it from insect pests and rotting. Some manufacturers offer guarantees of up to thirty years against insect

Above: Tree-house interior showing pull-up timber trapdoor.

Above: Tree-house interior showing deep shelf designed for storage, toy display and sleeping out.

attack or rotting. This system is often referred to as CCA (Copper Chrome Arsenic) protection.

Properly treated timber can be safely buried in the ground with no additional precautions. For basic posts and small-scale structures the posts may be set in holes dug in the ground. In these cases the backfilling will be sufficient to support the structure. For swings, which are subject to additional loadings, it is preferable to pour a concrete pad around the post to secure it in position.

The connections between the different parts of a play structure need to be designed for the severe outdoor exposure. Any steel parts should be hot dip galvanised and need to be well engineered to take account of the weather, and potential rusting. Bolt heads should be recessed to prevent injuries and provide further protection from the elements.

You may find it necessary to make or buy special connection pieces to suit your design. There are many small metalworking shops which are set up to do this type of work. Check with your local garage or building supplier. Some of the larger hardware shops offer the service of cutting steel plates and bar stock to size for their customers. Some will also send work away for galvanising. Some specialist galvanising companies are able to do smaller work, and it is worth checking the *Yellow Pages* for a suitable supplier.

Carpentry

Working with round timber is a little different to working with conventional square-cut timber. Accuracy is less critical and the techniques have their roots in the famous Australian bush carpentry. Almost all of the joints could be cut with a chainsaw! However, the more conventional carpentry tools will be sufficient. If you don't have a broad-bladed chisel (size 25–30 mm), this may be the right time to buy one. Otherwise, a rip saw and an electric drill, along with a few other basic tools, is all that you will need.

When you join two pieces of round section timber together, the joint has to be made to give two flat surfaces on each piece. With the two flat sections fitted together, the neatest and best way to hold them in position is to use bolts. As mentioned previously, the bolt heads should be set below the surface. This is done with a method called 'counterboring'. The hole through the timber is enlarged enough to take the washer and the nut to a depth of a little more than the combined height of the washer and the nut/bolt head.

To do this properly the length of the bolts used should be carefully selected. If two sections of timber, both 100 mm thick, are to be joined — and we remove 25 mm from each mating face — then a bolt of 150 mm will be perfect. The counterboring has to be done carefully so that the bolt-head does not disappear too deeply into the timber. Minor inaccuracies can be adjusted with a few extra washers.

Project: BUILDING A CUBBY-HOUSE

The elevated cubby-house shown here is just a starting point. All the basic joints that you will need for a whole variety of structures using round treated logs are also shown.

The dimensions of your structure will depend on the space available and the ages of your children. For an elevated cubby, leave enough space under it for the tallest kids. At least 1200 mm minimum clearance. Sometimes it's a good idea to make a sand pit under the cubby. The sand can help cushion any falls. Tan bark is also popular. If the kids are going to use the cubby frequently, any grass in the immediate surroundings will not survive, so some soft landscaping is necessary.

The posts for a cubby are set into concrete footings, but if the ground is firm enough you may want to simply use compacted earth. The trick is to prop the posts into position until the structure is complete before completing the footings. In this way you will be able to finally adjust the posts as required.

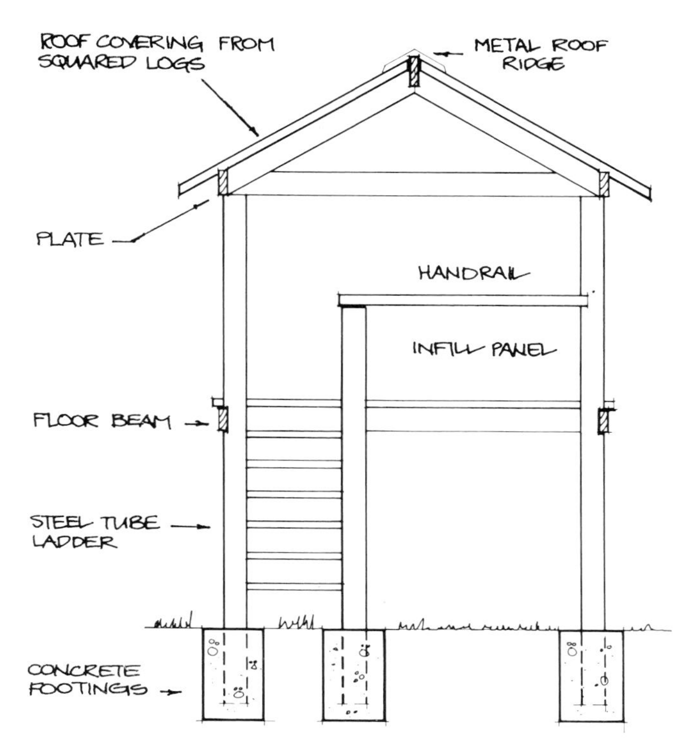

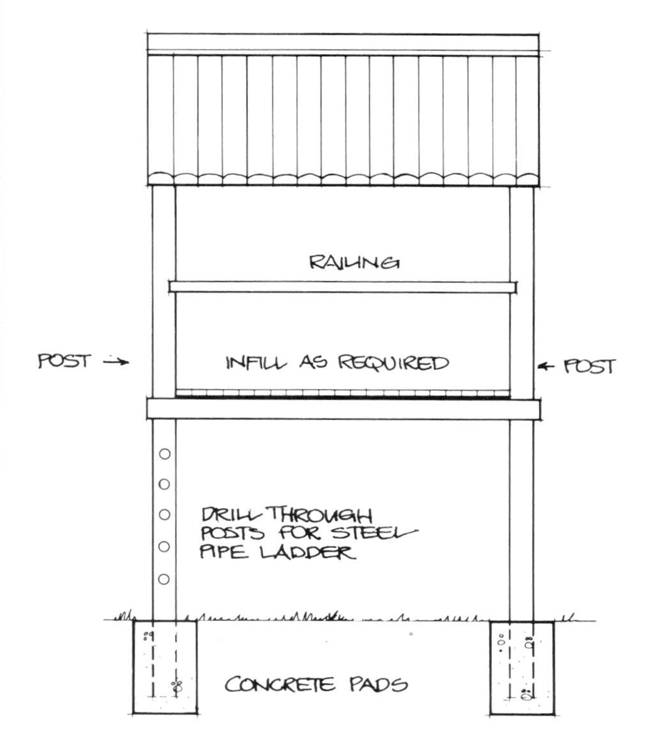

This cubby design uses a combination of round timber for the posts and roofing, and sawn rectangular timber for the deck and railings. For joining members together a variation of the housing joint is used. This is made by cutting two parallel saw cuts into the round log and removing a flat section. The rails and joists fit neatly into the recess and the connection is secured by bolting. As the timber treatment can affect steel fittings, all steel bolts and nails should be galvanised.

Ladders may be made by drilling 40 mm diameter holes through the posts at about 200 mm centres, and fitting lengths of galvanised steel waterpipe as the step treads.

The roof is made from cut sections of logs secured to the plates and at the vertex of the roof at the ridge. A metal flashing covers the ridge. This design is not wholly waterproof, but for an open cubby-house this is not a critical factor. For a waterproof roof, metal roofing or waterproof plywood can be used.

The sky is the limit with children's play structures, and the most successful ones are a product of creative design, usually involving the children in both design and construction.

JOINERY FOR LOG PROJECTS

Working with timber in the form of logs is no more difficult than with normal rectangular sections. The housing joint is commonly used as it gives more rigidity in the joint. Make sure that the joint is a good tight fit as the timber will tend to shrink with time. When working by yourself, the other advantage of housed joints is that the joint is self-supporting until the bolts are fixed in position.

A 90 degree cross joint is marked out by placing the members in position and indicating the point of intersection. Make two saw cuts inside the lines and use a broad chisel to remove the housing. Make sure that the bottom of your cut is properly aligned and doesn't cause twisting. Remove more material if necessary to get a perfect fit, but work carefully, checking as you go to ensure accurate work. Galvanised steel bolts are used to secure joints.

A combination of round logs and treated rectangular sections is sometimes useful to simplify the connections between members. The floor joists and bearers for a cubby-house are best made from squared timber, but the posts and structure can be made from logs.

Dowelled joints, where the two pieces are secured in position with round timber dowels, are sometimes used in log structures. In the past, when all work was done with round logs and nails were very expensive, bush carpenters often had to use elaborate joints to save using nails.

As a general rule, it's best to recess the heads of all metal bolts as a safety measure. First, use a small diameter drill to make a pilot hole; then use a 'counterbore' to drill a recess to the diameter of the bolthead (of sufficient depth to conceal the bolthead). Finish off the job by drilling the clearance hole for the shank of the bolts.

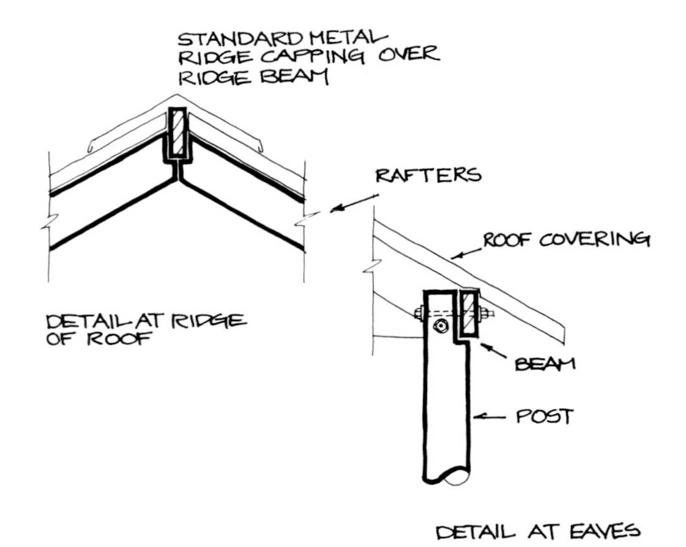

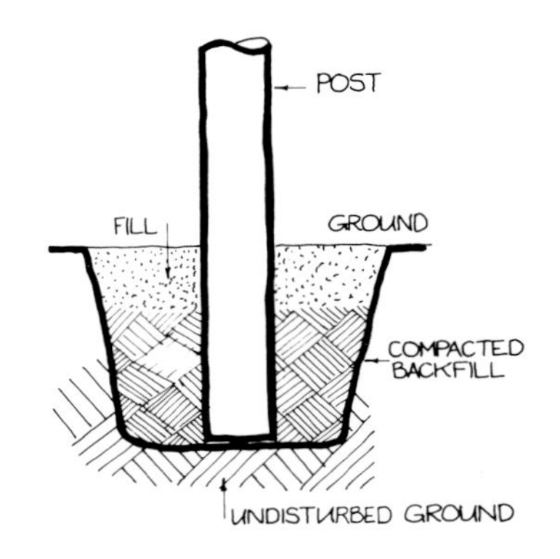

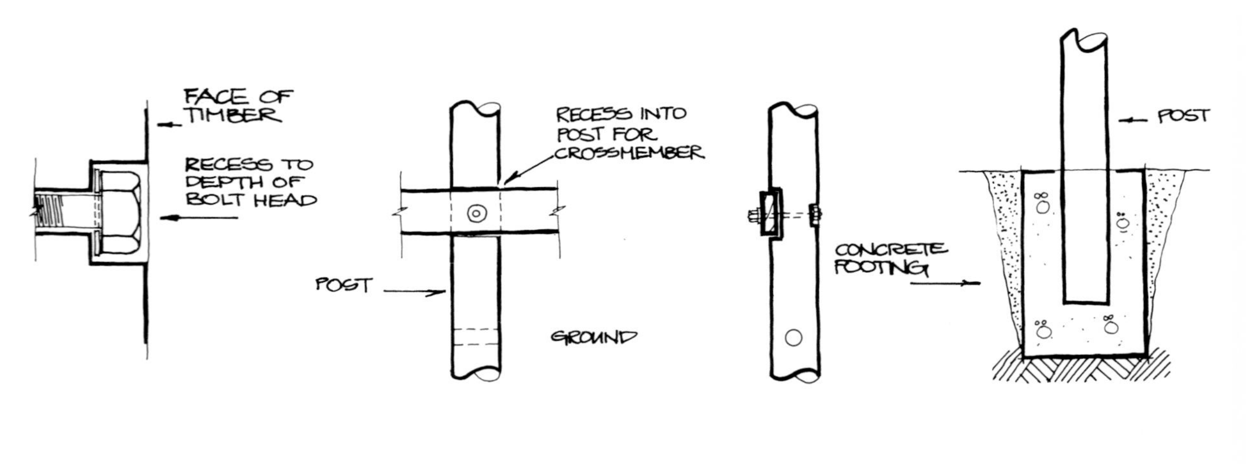

A GUIDE TO CONCRETE

Concrete is a mixture of portland cement powder, sand and an aggregate (crushed rock) in carefully measured proportions. By a complex series of chemical reactions, it turns from a wet mixture into the rock-like substance we know as concrete.

A suitable mix for small raft footings and slabs would be a 4:2:1 ratio, by volume, of aggregate, sand and portland cement. Up to 50 per cent of the cement content can be substituted by hydrated lime for non-critical areas. Lime gives 'work-ability' to the mix.

Concrete is mixed by hand or in a concrete mixer. For larger jobs, it is brought to the site by special Ready-Mix trucks. You can obtain amounts as small as one half of one cubic metre from some suppliers, who have smaller versions of these trucks.

Builders and concrete suppliers vary the proportions of the concrete according to job requirements. Other materials are often added to further control the properties of the concrete. There are even special concrete mixes which are designed to set under water. The technical property of the concrete mix which most affects the home-owner is the design strength of the concrete. Design strength is measured in 'mega-pascals', and the usual abbreviation is MPa.

For general footing work and the construction of light duty concrete driveways, the usual grade of concrete specified is '20 MPa'. If you are in any doubt about the kind of concrete to use for one of your projects, consult a concrete supplier. If the design is critical, consult a qualified structural engineer.

Concrete Slabs

Most homes have some form of concrete slab in the garden, whether it's a small raft footing for a barbecue, a side path or a large and extensive driveway.

Laying a concrete slab is not difficult provided a few simple rules are followed. The simplest one is not to take on a large area of concreting unless you are reasonably fit. Concrete weighs about 2500 kilograms per cubic metre, and the work can be back-breaking. You will be able to do the job far more easily if you have help.

Formwork

For all concreting work some kind of 'formwork' is required to retain the wet mix in position until it hardens. In paths and slabs the formwork consists of the levelled foundation, often covered with a plastic sheet, and the 'edge boards'. The purpose of plastic concrete underlay is to prevent the concrete mix from drying out too rapidly due to water loss. It also provides a vapour barrier to the passage of moisture after the concrete has set.

Edge boards are used to form the edges of a concrete slab, and are normally made from strong timber. Straight-grained 100 mm × 50 mm timber is suitable for domestic paths. The edge boards are kept in position by stakes driven into the ground.

Reinforcement

An essential component of concrete is the steel reinforcement. Concrete by itself is very strong in compression, but very weak in tension. Steel rods, on the other hand are extremely strong in tension, but weak in compression. Concrete and reinforcing steel together provide the ideal combination of properties.

For general work around the home, prefabricated steel mesh is often the most efficient solution for reinforcement. Reinforcement mesh is made from steel wire or rod welded together at right angles. It provides optimum reinforcement to concrete at the lowest possible cost. For paths and paving slabs which don't need to support substantial loads, 'path mesh' or 'F-72' should be adequate.

Most larger building suppliers stock the standard sizes of reinforcement mesh and trench mesh (a special grade designed for concrete footings), and will cut it to suit your requirements, as well as delivering it to site.

Concrete Finishing

The small flat piece of timber with a wooden handle used for finishing off concrete is called a wood float. It gives a relatively smooth finish, but still leaves some grip to stop the path surface from becoming slippery. The steel trowel, which looks similar to the wood float, gives a smooth, glass-like finish. This is very useful for driveway areas as it makes cleaning easier. Steel trowel finishes are unsuitable for paths for safety reasons.

Where traction and grip are important, a 'broomed' finish is often used. This is achieved by waiting until the concrete has reached its initial set, and then drawing a stiff brush over the surface. The edges and joints in the path are normally 'tooled' with a jointing tool. There are several types of jointing tools designed for edges and joints. Using these will help you to achieve a professional result.

Concrete slabs can be coloured and embossed to resemble cobblestones and other unit paving styles. In this process the concrete is laid as normal, but the mix has special colouring agents added. A mould is pressed into the wet surface of the concrete to form the required texture. The effect is an attractive and highly durable surface without some of the problems of settlement that plague unit paving. The concrete can also be designed to take the heaviest of loads.

WHERE TO GO FOR ADVICE AND INFORMATION

New South Wales

Building Information Centre
525 Elizabeth Street
South Sydney 2012

Building Services Corporation
(formerly the Builder's Licensing Board)
6-12 Atchison St
St Leonards 2065

Cement and Concrete Association
100 Walker St
North Sydney 2060

Radiata Pine Association
10 Pitt St
Parramatta 2150

Royal Australian Institute of Architects
3 Manning St
Potts Point 2011

Standards Association of Australia
Standards House
80 Arthur Street
North Sydney 2060

Timber Development Association
525 Elizabeth Street
South Sydney 2012

Victoria

Building Development and Display
332 Albert Street East
Melbourne 3000

Royal Australian Institute of Architects
30 Howe Crescent
South Melbourne 3205

Standards Association of Australia
Clunies Ross House
191 Royal Parade
Parkville 3052

Timber Merchants' Association
184 Whitehorse Road
Blackburn 3130

Queensland

Royal Australian Institute of Architects
Cnr Mary and Albert Streets
Brisbane 4000

Standards Association of Australia
447 Upper Edward Street
Brisbane 4000

Timber Advisory Bureau
5 Dunlop Street
Newstead 4006

South Australia

Building and Home Improvements Centre
113 Anzac Highway
Ashford 5035

Royal Australian Institute of Architects
GPO Box 2438
Adelaide 5001

Standards Association of Australia
11 Bagot Street
North Adelaide 5006

Timber Development Association of South Australia
113 Anzac Highway
Ashford 5035

Western Australia

MBA Building Information Centre
161 Havelock Street
West Perth 6005

Royal Australian Institute of Architects
PO Box 191
West Perth 6005

Forest Products Association of Western Australia
103 Collins Street
West Perth 6005

Standards Association of Australia
11–13 Lucknow Place
West Perth 6005

Tasmania

Royal Australian Institute of Architects
GPO Box 1139L
Manuka 2603

Standards Association of Australia
97 Murray Street
Hobart 7006

Tasmania Timber Promotion Board
68 York Street
Launceston 7250

Northern Territory

Royal Australian Institute of Architects
PO Box 1017
Darwin 5794

Standards Association of Australia
c/- Master Builders Association
191 Stuart Highway
Darwin 5790

Australian Capital Territory

Royal Australian Institute of Architects
PO Box 99
Manuka 2603

GLOSSARY OF BUILDING TERMS

AGGREGATE Crushed stone screened to a size suitable for concrete. The maximum size of stone, suitable for general use is 20 mm. Aggregate is obtained from better builders' suppliers. It is sometimes called bluemetal or simply metal. Bluemetal is crushed basalt rock.

ALTITUDE (Astronomical) The vertical angle between the sun and the horizon. The maximum and minimum altitudes of the sun are important factors in the design of solar pergolas.

ARCHBAR A steel structural beam designed to support brickwork over an opening such as a window or doorway. A common size of archbar is 75 mm × 10 mm flat bar. Archbars are commonly galvanised and are supplied in various lengths, according to customers' orders.

ARRIS The 90 degree corner of a piece of timber. The term is used as a substitute for 'edge'.

BALUSTER These are the individual members of a balustrade.

BALUSTRADE A railing usually around a deck or balcony. They may be constructed from timber or metal, but for the average handyman, timber is worked more easily.

BATTENS In general terms, a piece of timber, having a small dimension and no major structural purpose. Battens are used to support roof tiles and for shading on pergolas and are usually about 50 mm x 25 mm.

BATTER The slope of a retaining wall away from the vertical, towards the retained material. For low walls a batter of about one in ten is normally specified.

BEARER In floor construction, the structural member that carries the load of the joists to the structure. Bearers are usually fixed to posts, stumps or to brick piers.

BLEEDING A feature of unseasoned or green timber that will cause staining of concrete until all the sap leaches out of the new timber. More an unsightly nuisance than a serious defect, it may be minimised by buying well-seasoned timber or secondhand demolition timber.

BLINDING A levelling bed of sand or other fine grain material usually laid under paving.

BOLSTER A short-handled hammer with a heavy square head used for cutting bricks etc. Sometimes also called a lump hammer or a lumpy.

BOND The way bricks are laid in a wall or paving so that they interlock. The simplest is called 'stretcher bond'. This is laid by placing each successive course, or layer, so that each joint is staggered by one half brick length.

BRICK SETT A heavy, broad-bladed, cold chisel used by masons and bricklayers for brick cutting.

BUILDERS' SAND A clean, beach sand that has been washed to remove all salt and organic matter. Available from all builders' suppliers, either in bulk or bagged.

BUILDING APPLICATION (B.A.) Or sometimes Building Approval. ('Have you got the B.A. yet?'). Most building works must be approved by your local council prior to commencement. The building application is a formal request for permission to build, and will require specially prepared plans and specifications.

BUSH SAND An alternative to builders' sand but only suitable for mortars and not for rendering work. Bush sand gives a fatty mortar that is easy to work.

CCA TREATMENT Pressure preservative treatment for timber using Copper, Cadmium and Arsenic. Gives timber a characteristic greenish grey colour and provides long-term protection from insect attack.

CEMENT Portland cement is used as a basis for all bricklaying mortars and in concrete. One bag weighs 50 kg which means there are 20 bags to 1 tonne. Cement should be purchased in sufficient quantities for the job at hand; it tends to harden with atmospheric moisture.

CEMENT MORTAR A general purpose mortar for all bricklaying. It comprises one part Portland cement and three parts sand. This mortar gives a grey to off white joint.

CHAIR A device designed to support the reinforcing steel and maintain it in the desired position during the pouring of concrete. Chairs are used in conjunction with pressed metal pans to prevent the plastic underlay from being damaged.

CHAMFER The process of removing the edges or arrises of a length of timber, usually by planing at an angle of 45 degrees, to a small depth.

CHUCK The part of a drill or other tool which holds a removable cutting implement. In the case of a drill, the 'bit'; similarly a 'router' has a chuck for the 'cutters'. Chucks are operated with a 'chuck key'.

COMPO MORTAR A bricklaying mortar which comprises one part hydrated lime, one part Portland cement and six parts sand. A weaker mortar than 3:1 cement mortar but more economical and simpler to work.

CONCRETE A mixture of cement, sand and aggregate mixed in varying proportions, according to the strength and purpose required. For footing and sundry paving work, a mix of four parts aggregate, two parts sand and one part Portland cement is satisfactory.

CONCRETE COVER The thickness of concrete covering the reinforcement mesh. Adequate concrete cover is essential to prevent 'concrete cancer' which is, simply, the reinforcement steel rusting.

CORED BRICKS An alternative to a frog in a brick. Usually found in extruded bricks, the cored out holes reduce the weight of the brick and perform the same function as the frog.

COUNTERBORING A method of carpentry which conceals the heads of bolts and nuts in posts and other timber construction. The centre of the intended bolts is marked and the first drilling is made with a boring bit, of sufficient diameter to clear the washer and deep enough to hide the bolt head. The clearance hole for the bolt shank is then drilled through. To drill the counterbore where the drill has emerged, a piece of scrap timber may be clamped over the hole and the centre carefully marked. When drilling such a hole, constantly check the depth to ensure that the counterbore is correct.

COUNTERSUNK A special drill bit is used to enlarge a screw hole so that the screw head will sit below or flush with the surface.

COURSE In all masonry work, a course is a rise in level of one unit.

DAR Dressed all round. Rough sawn timber ordered as DAR will be planed smooth and square on all four faces.

DECKING The actual walking surface of a timber deck. Usually made from dressed and pretreated hardwoods like tallowwood.

DEVELOPMENT APPLICATION (D.A.) An application to your local council for permission to erect a structure which requires planning approval. This is often necessary when you are enlarging your home, or building a new one. For minor works a Building Application (B.A.) is normally all that is required.

FINIAL The topmost part of a structure. In a gazebo, it could also be called a 'hub'.

FOOTING The lowest part of a building or structure that rests on the ground. Usually constructed from reinforced concrete or, in small structures, of bricks.

FORMWORK A timber frame made out of stakes driven into the ground and perimeter boards that sets out the level of the top edge of a concrete slab. Formwork should be well constructed so that it will not move during the concrete pour, but should also be designed that it may be easily removed when the concrete has set.

FOUNDATION The ground on which the footing is built. Sometimes used interchangeably but incorrectly with footings. This term relates to the ground not the building.

FROG A recess pressed into a clay brick before firing. The frog assists the levelling of the brick courses and strengthens the wall. (See also Cored Bricks.)

Ga (for gauge) A method for measuring thickness of sheet metal, and for the diameters of some fittings like screws and nails.

GALVANISING An electrolytic process that coats raw steel with a protective layer of zinc. This process is recommended for all exposed external steel work. Hot dip galvanising is the most effective process and users should specify this type of service.

GROUTS Fine cement pastes used for filling gaps, particularly in ceramic tiling.

HEADER BRICK, HEADER COURSE A header is a brick that is laid across the line of the wall of brickwork. A brick on edge header is a header course laid with bricks on edge rather than on the flat and is a useful finish to a wall.

HOUSING JOINT In carpentry, a joint where one or several components are fitted, with a portion of each piece being removed to allow them to 'marry'. A housing joint is often used to connect joists or bearers to posts. The advantage is that the housing provides temporary support during construction until the connections are completed.

JOIST A horizontal structural member, supporting a floor, deck or ceiling. The joists are fixed to plates in the case of ceiling joists or to bearers in the case of a floor. In pergolas, sometimes the terms joists and rafters are used interchangeably.

KNEE BRACE A short, diagonal bracing member usually applied to the bracing of a pergola to post connection. Bracing is essential to stop a pergola swaying.

LARRY A long-handled hoe with a hole in the blade. It is most useful for mixing mortar in a wheelbarrow.

LIME Hydrated builders' lime is supplied in 50 kg bags and is a useful additive for all bricklaying mortars as an aid to workability. (See also Compo Mortar.)

LEVEL Quite obviously, this means perfectly horizontal. If in doubt, check your construction several times with a long, spirit level, reversing the level to eliminate errors.

MERCHANTABLE GRADE TIMBER A lower quality grade of timber that will suffice for all but the most critical applications. This grade should be specified for most jobs.

MORTISE AND TENON A mortise is a rectangular hole cut into one piece of timber. A tenon is a tongue-shaped section designed to fit into a mortise. Thus mortise and tenon joint. The haunched mortise and tenon joint is a variant of the mortise and tenon joint, mostly used for cabinet work.

PATH MESH A grade of pre-fabricated steel reinforcement designed for light paving work.

PERGOLA An unroofed or partially-roofed frame, designed to provide a base for climbing plants and to give quasi-shelter. Pergolas provide a transition space from the inside of the house to the outside and are an effective part of an outdoor living space. Pergolas are occasionally freestanding, but are usually attached to the house.

PERPENDS The vertical joints between bricks.

PLATE When applied to timber framing, this term means a horizontal load bearing member. A bottom plate may be found at the base of a stud wall and a top plate will be at the top of the wall, supporting the ceiling joists and roof structure. A wall plate is fixed to a wall to pick up the rafters of a pergola or other structure.

PLUMB A builders' term meaning perfectly vertical or perpendicular.

RAFTER An angled, supporting member, usually in a roof. In a pergola where the angle or slope of the rafters is small, the rafter may be confused with a joist.

REBATE A rebate is a recess or step, usually of rectangular section, cut into a surface or along the edge of a piece of timber to receive a mating piece. Sometimes pronounced 'rabbet'. A router is particularly useful for rebate work.

REINFORCEMENT REBAR Steel bar stock and prefabricated mesh used to strengthen concrete work. For complicated work where suspended slabs or heavy loads will be supported, the advice of a qualified structural engineer will be necessary for the reinforcement steel design. For minor building works, such as paths and barbecue footings, the use of path mesh should suffice.

RETEMPER Adding additional water to a setting cement mix. Not recommended as it results in a weakened mix.

ROUTER A very useful power tool, which has a vertical spindle and chuck with a high speed cutter designed to cut recesses of an almost limitless variety.

SCREED A thin layer, usually of mortar, used as a bed for tiles or a finishing application to rough concrete.

SELECT GRADE In timber, a superior grade that costs more but allows the purchaser to choose the best piece of timber for a specific purpose.

SHADECLOTHS Woven fabrics usually made from man-made fibres, either of split tape or filament. Most have especially good resistance to sunlight and UV degradation. Shadecloths are available in several grades which are specified by their light transmittance. A 70 per cent shadecloth will block out 70 per cent of the light. For pergola and shadehouse use, grades of 80 to 90 per cent are suggested.

SHOE OR CONNECTOR A prefabricated metal connection device, specially designed to fit the bases of timber posts and connect into concrete slabs and footings. Many styles are available ex-stock from hardware suppliers, and may be made to suit a particular application by a well equipped metal worker.

SKEW-NAILED When a nail is driven in at an angle. Rafters are connected to plates with a skew nail driven through the bottom section of both faces of the rafter into the plate below.

SOLDIER COURSE A course of bricks, often used as a finish to a wall, laid with the bricks vertical. A very elegant detail if well executed.

SPARROW PICK A technique which uses a sharp-pointed pick to roughen a concrete surface in order to obtain a better surface for other materials to adhere to.

STOP CHAMFERING The planing of the edge of a piece of timber is called chamfering. When the chamfer is stopped short of the ends or connections, it is called stop chamfering. Chamfering is desirable to reduce splitting and splintering and gives a neat, professional finish. Stop chamfering is often used in traditional timberwork, especially in verandahs.

STRETCHER BOND Common brickwork where the vertical joints, called perpends are staggered by one half brick on each successive course.

STRINGER The angled beam that supports the treads on a stair.

STUD
STUD FRAMING The vertical timber of a wall frame. Usually the frame is from 75 mm x 50 mm or 100 mm x 50 mm timber and the studs are housed into the plates. Locating the studs in an existing wall is essential for the fixing of a wall plate for a pergola.

TENON SAW A short, rigid-bladed saw with a steel stiffener to keep the blade dead straight. Used for critical work when cutting joints.

TOLERANCE An allowance for reasonable accuracy.

WALL ANCHORS Proprietary fittings that fit into pre-drilled holes in masonry and by expansion, secure varying items. A vast range of bolts, inserts and studs is available to the home handyman. Ramset and Dynabolt are two popular trade names of these all-purpose designed fixings.